SUPERCOMPUTERS IN
THEORETICAL AND
EXPERIMENTAL SCIENCE

Supercomputers in Theoretical and Experimental Science

Edited by

Jozef T. Devreese

and

Piet Van Camp

University of Antwerp
Antwerp, Belgium

Plenum Press • New York and London

Library of Congress Cataloging in Publication Data

International Workshop on the Use of Supercomputers in Theoretical Science (1984:
  Corsendonk Conference Center)
  Supercomputers in theoretical and experimental science.

"Proceedings of the International Workshop on the Use of Supercomputers in
Theoretical Science, held July 30–August 1, 1984, at the Conference Center, Priorij,
Belgium"–T.p. verso.
  Includes bibliographies and index.
  1. Supercomputers–Congresses. I. Devreese, J. T. (Jozef T.) II. Van Camp, P. E.
(Piet E.)
QA76.5.I623 1984   004.1'1   85-19331
ISBN-13: 978-1-4684-5023-1   e-ISBN-13: 978-1-4684-5021-7
DOI: 10.1007/978-1-4684-5021-7

Proceedings of the International Workshop on the Use of Supercomputers
in Theoretical Science, held July 30–August 1, 1984, at the Conference Center,
Priorij, Corsendonk, Belgium

©1985 Plenum Press, New York
Softcover reprint of the hardcover 1st edition 1985
A Division of Plenum Publishing Corporation
233 Spring Street, New York, N.Y. 10013

PREFACE

The International Workshop on "The Use of Supercomputers in Theoretical Science" took place from July 30
till August 1, 1984, at the Conference Center of the
"Priorij Corsendonk", close to the city of Antwerpen,
Belgium.

During the past decade computational science has
developed itself to a third methodology besides the
experimental and theoretical sciences. This remarkable
evolution was only possible due to a drastic increase of
the computational power of present day computers. Indeed,
computational physics and chemistry as such is certainly
not new, but it was only during the past ten years or so
that realistic problems could be solved numerically to a
sufficient degree of accuracy. During this workshop the
state-of-the-art in high speed computation was presented
by a team of lecturers who are well known for their
competence in this field.

It is a pleasure to thank several organizations and
companies who made this workshop possible. First of all,
the main sponsors: the Belgian National Science Foundation (NFWO-FNRS) and the "Universitaire Instelling Antwerpen". Next, the co-sponsors: Agfa-Gevaert N.V.,
Control Data Belgium and the Belgian Ministry of Education.

Special thanks are due to Dr. P.E. Van Camp and Drs.
H. Nachtegaele for the practical organization of this
workshop. I would also like to thank Mrs. H. Evans for
typing the manuscripts and for preparing the author and
subject index.

Last but not least I express my gratitude to Mr. D. Van Der Brempt, manager of the Corsendonk Conference Center, for the way he and his staff took care of the local organization at the Conference Center.

J.T. Devreese
Professor of Theoretical Physics

June 1985

CONTENTS

I. INTRODUCTION

II. SIMD SUPERCOMPUTERS

III. MIMD SUPERCOMPUTERS

IV. APPLICATIONS

V. INDEXES

I. INTRODUCTION

SCIENCE, SIMULATION AND SUPERCOMPUTERS

W.C. Nieuwpoort

Laboratory of Chemical Physics and Material
Science Centre
University of Groningen, The Netherlands

INTRODUCTION

Contributing to the opening session of this seminar on supercomputers and their uses is, I must confess, less easy than I thought it to be when Professor Devreese asked me to do this during last week's Institute on the Theory of Condensed Matter Physics. Whatever I could say technically on the subject would not only probably pale in the light of the assembled expertise here, but would most certainly be utterly boring for our guests who have been invited to attend just this opening. What I am going to say therefore represents a compromise with a distinct bias towards this latter group. The theme of my talk derives from a remark made by Professor V. Heine at the beginning of last week's Institute on the subject of computation in physics and a seemingly conflicting statement on the same subject that I have read in a recent issue of the journal Physics Today of the American Institute of Physics.

The remark of Professor Heine [1] was actually a complaint about the often reluctant, if not derogatory, attitude of theoretical physicists towards large scale computational methods. An attitude taken in spite of the fact that the development of such methods, the design and construction of the associated program systems and their intelligent use, require a combination of theoretical knowledge and experimental skill that is quite comparable with that needed in more traditional research activities in physics (and chemistry I should add). Such

3

negative viewpoints are of course of little help when trying to convince funding agencies of the importance of supercomputer facilities for the progress of basic research in physics.

A FAR-REACHING MANIFESTO

Quite a different attitude is expressed in a rather sweeping declaration, a manifesto even, that I ran across recently in a computer-dedicated issue of Physics Today [2]. This manifesto was presented after the first meeting in January this year of the NSF Advisory Committee for Advanced Scientific Computing by its chairman and is quoted under the heading "Not since Galileo":
> "Science is undergoing a structural transition from two broad methodologies to three - namely from experimental and theoretical science to include tne additional category of computation- al and information science. A comparable example of such change occurred with the development of systematic experimental science at the time of Galileo".

It goes on by pointing out the extreme importance of this transition for science and engineering, for universities and industries, not only because of its impact on research and design methodology but also on training and educational programs.

I have no difficulty at all in sharing the enthusiasm and the strong belief in the importance of computational methods expressed in this declaration. I have some doubts, however, about the suggestion that we are witness- ing a methodological development that is new in principle: the development of computational science as a new category besides experimental and theoretical science, so well established in western society since, indeed, the time of Galileo. In fact I think that the manifesto overstates the case of computational science in this respect. While this may be good for some purposes, it can only add to the arguments underlying the negative attitudes referred to above and why should one do that. I shall try to clarify this (mild) objection by remind- ing you briefly of what is involved in what we call the scientific process and by drawing attention to the essential rôle that computations, simple or complex, have always played in this process.

THE SCIENTIFIC PROCESS

Science, one might say, consists of loops that start and end with observations, with the theoretical process of logical reasoning, inductively and deductively, lying in between. The diagram below describes this schematically. Depending on the field of interest sets of observations are abstracted from the real world. The ever curious and inquisitive human mind notices regularities and establishes relations between observations. Underlying "physical" models are conceived that, to be operationally useful, must be translated into "mathematical" models: a representation of the real world which allows logical deductions to be made. These deductions must be confronted with the original observations in order to verify (or falsify if one wishes) the theory and, most importantly, they can lead to predictions about observations that are not yet made and hence to new experiments. When the predictive power of a theory has reached a certain confidence level a gradual shift of its importance from science to engineering may take place. We note that theory also allows us to carry out "gedanken" or "thought" experiments for which Einstein was so famous for. These are experiments that are logically possible but are too difficult or even impossible to carry out in practice. They are important for instance to understand the limits of validity of a theory and for improving our understanding and physical insight in general. This brief overview of what is involved in the scientific method will suffice, I think, to show that computation in one form or another is, and always has been, an integral part of the process just as the categories of observation and theoretical model forming. I have emphasized this at the right of the diagram by applying a further reduction, where all activities on

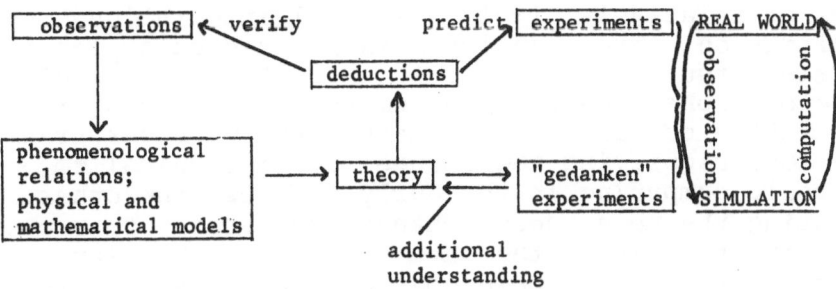

the theoretical level are seen as efforts to obtain consistent simulation models of the real world. Gedanken experiments being absorbed in the more modern term computer experiments. Observation and computation representing the necessary input and output communication channels between the real world and its simulated counterpart, no truly scientific theory being possible without them. Just like any other part of the scientific process, the computational one requires inspiration and creative moments of its practioners, and there is no basis, apart from personal taste, for attributing to it a priori a lower intrinsic value. Such prejudice may in fact, as all prejudices, be quite harmful to scientific progress.

A REVOLUTION IN SCALE

While computation conceptually represents nothing new in the methodology of science, the means to do it have, as we all know, radically changed in the last three decades or so. And, of course, this change and its consequences are what lies behind declarations like the manifesto quoted above. The change in scale of the computational power available to us has been truly remarkable and shows revolutionary characteristics. I think that in any field of application examples can be found to illustrate this. Let me give an example from atomic physics, a celebrated self-consistent field calculation including exchange on the copper atom, by Hartree and collaborators in the late thirties at Cambridge. The reason for choosing it is that an estimate exists about the man-years needed to carry out the original calculation and that essentially the same calculations have been repeated many times since electronic computation became possible. It can serve therefore as a convenient unit of computational power P, which is defined as the reciprocal of the time needed to complete the calculation. The calculation involves an iterative series of relatively simple numerical integrations in one dimension. The quantity P is plotted on a logarithmic scale versus time in Fig. 1 (1938, $P \approx 5.10^{-8} sec^{-1}$). The revolutionary development emerging from this figure needs little comment. It is marked by two clear transitions. One signifying the change from mechanical to electronic computing in the early fifties, the other indicating the more recent change from scalar to vector processing systems in the late seventies. The large increase in processing speed goes of necessity hand in hand with the availability of large main memories and high rates of data transport to and from peripheral memory

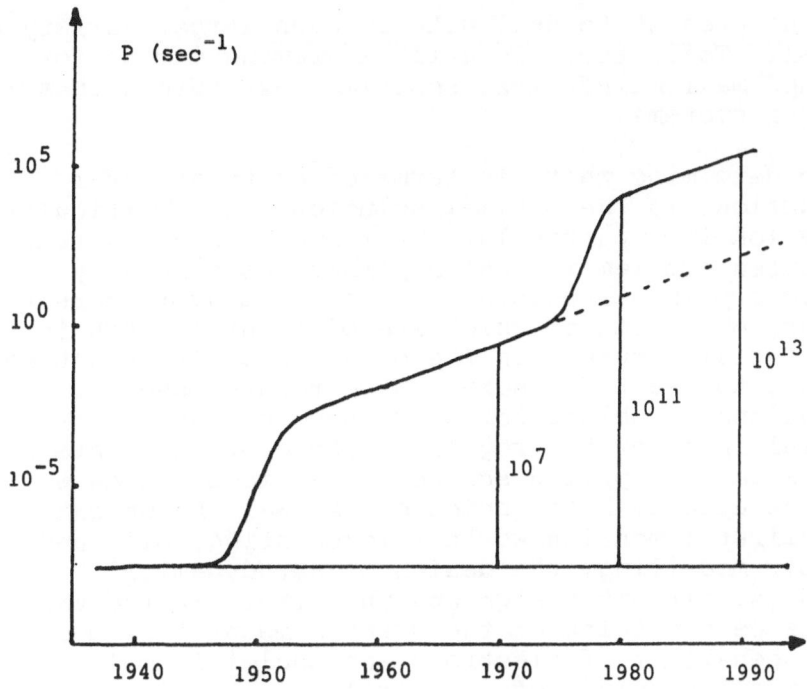

Fig. 1. Development of computational power.

devices. Today's supercomputers, vector processors like
CRAY and CYBER 205, can perform at maximum rates of a
few hundred MFLOPS (million floating point operations
per second) and have memories of 1-4 Mwds of 64 bits per
word. In a year or so we may see increases towards
GFLOP in speed and 4-32 Mwds in memory and towards the
end of the decade machines of 10 GFLOP maximum rate and
32-256 Mwds of memory are to be expected.

SUPERCOMPUTERS, WHO NEEDS THEM?

Supercomputers by definition are the fastest and
the biggest. And in theory, a well-balanced configuration, run with utmost efficiency and to full capacity,
can be the cheapest as well per unit of computation, say
a MFLOP. To conclude from this that supercomputers offer
a cost-effective solution to most if not all computational problems would be a grave mistake, however. This is
amply demonstrated by the large variety of hardware
solutions that have been chosen in the past and that are

chosen at present to deal with an even larger variety of problems. Today there is still a growing demand for mini- and maxi-mainframes, sometimes extended with array processor systems.

To determine what, in terms of costs <u>and</u> user satisfaction, is the optimal solution for a particular application in a particular environment is an extremely complicated problem for which perhaps no generally satisfactory answer exists. There are many factors to consider, only some of which are of a purely technical and financial nature. On the other hand, it is not so difficult to see that there are certain classes of computational problems for which supercomputers are essential in order to progress. These problems are found in the areas of applied science or engineering as well as in the area of basic science. As well-known examples of the first I mention weather forecasting, oil- and gas reservoir modelling, the design of aerodynamical structures, circuit design and the like. In the basic sciences we can think of the quantum mechanical models for the behaviour of electrons and nuclei in molecules and solids, the study of classical many-particle systems by Monte-Carlo or molecular dynamics methods, many-particle scattering problems etc. etc. All these problems involve complex mathematical models with many degrees of freedom whose evolution in time and space must be explicitly treated. These models in turn lead to large non-decomposable computer programs that must be run many times over and over again with heavy demands on all system resources.

Nowadays the size of a problem by itself does not require the use of a supercomputer, although it often can be justified on a price/performance basis. What makes a supercomputer necessary is the fact that results, to be useful, must be obtainable within a certain critical time-span that is determined by the application at hand. This is obvious for the applied science and engineering examples given above. But also in the basic sciences a fruitful interaction between theory and experiment is most likely to occur when theoretical results can be obtained at rates that are comparable with those at which experimental results can be produced.

SOME FINAL REMARKS

Since time is running out let me finish with making some brief final statements:

- we have only seen the beginning of the possibilities of truly parallel processing.
- in order to make full use of the possibilities to come, experience and education are essential. This means that parallel processing systems must be accessible for scientists and students, but also that one must learn to rethink a simulation problem in the light of the new possibilities. Not only on the semantic level, but perhaps also at the level of the physical theory to be employed.
- in doing this it would be of great help if local front-ends were available on which vector programs could be fully simulated before sending them to a supercomputer for production.
- the computational sciences are in full swing and we are witnessing exciting developments. However, now and then we might heed the words of the old pessimist Schopenhauer [3], who said about learned men that read (not compute) a lot: "He who reads very much and almost the whole day, loses the ability to think for himself. Like someone who always rides forgets how to walk. This now is true for very many learned men: they have read themselves out of their wits!"

REFERENCES

[1] V. Heine, "Computation of electronic structure: its role in the development of solid state physics", opening lecture at the International Advanced Study Institute on Electronic Structure, Dynamics and Quantum Structural Properties of Condensed Matter, University of Antwerpen, Belgium, July 1984.
[2] Physics Today 37, May 1984, p. 61.
[3] A. Schopenhauer, "Die Welt als Wille und Vorstellung" (1819).

II. SIMD SUPERCOMPUTERS

VECTOR PROCESSING ON CRAY-1 AND CRAY X-MP

Ulrich Detert

Zentralinstitut für Angewandte Mathematik

Kernforschungsanlage Jülich GmbH, 5170 Jülich

1. INTRODUCTION

Vector processors provide great performance improvements in scientific and engineering computations. This is achieved by a specialized architecture designed for the fast processing of certain types of operations. On the other hand this specialization in hardware and architecture leads to a corresponding specialization in programming and program optimization for such machines.

This paper presents some aspects of vector processing on CRAY-1 and CRAY X-MP. First, principles of the CRAY-1 architecture will be sescribe. Differences between CRAY-1 and CRAY X-MP and their influence on performance will be pointed out. This will be done by the analysis of some FORTRAN kernels.

In the second part some properties of the CRAY FORTRAN compiler will be presented. The necessity of adequate program optimization will be pointed out.

2. PRINCIPLES OF CRAY-1 ARCHITECTURE

The CRAY-1 is a register oriented vector processor [1]. On its way from memory to the arithmetic and logical processing units (functional units) all data processed in either vector or scalar mode is passed through a set of vector or scalar registers, respectively. Vector registers are prepared to hold up to 64

vector elements, thus providing the pipelined vector
functional units with a contiguous stream of operands or
taking up the results of pipelined vector operations
(figure 1).

By this, the vector registers serve in a way as a
"data cache" between memory and functional units. As a
consequence, the "start up time", i.e. the time required
for the initialization of vector operations is relatively short. Vectorization of operations with vector lengths
greater than four or five pays off compared with the same
operations processed in scalar mode. On the other hand
a new vector instruction must be issued for each partial
vector of length 64 if the original vector to be processed contains more than 64 elements. Figure 2 shows the
typical execution time of a vector operation on CRAY-1
related to vector length.

As all data to be processed is passed through
registers, the performance of the CRAY is much dependent on the speed of the data transfer between memory and
registers. The CRAY-1 contains only one memory port
for this transfer (figure 1). As a consequence, vectors
(with up to 64 vector elements) have to be loaded and
stored sequentially one after the other.

In a typical example

        DO 1 I = 1,100
    1   C(I) = A(I) + B(I)

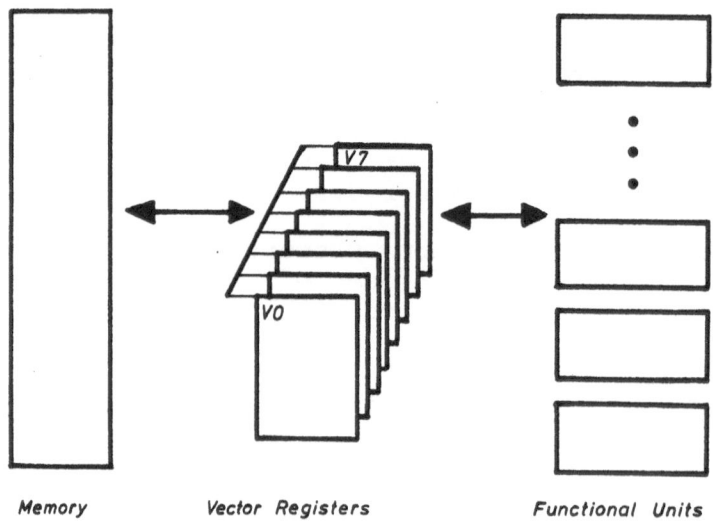

Memory   Vector Registers   Functional Units

Fig. 1. Data paths for CRAY-1 CPU.

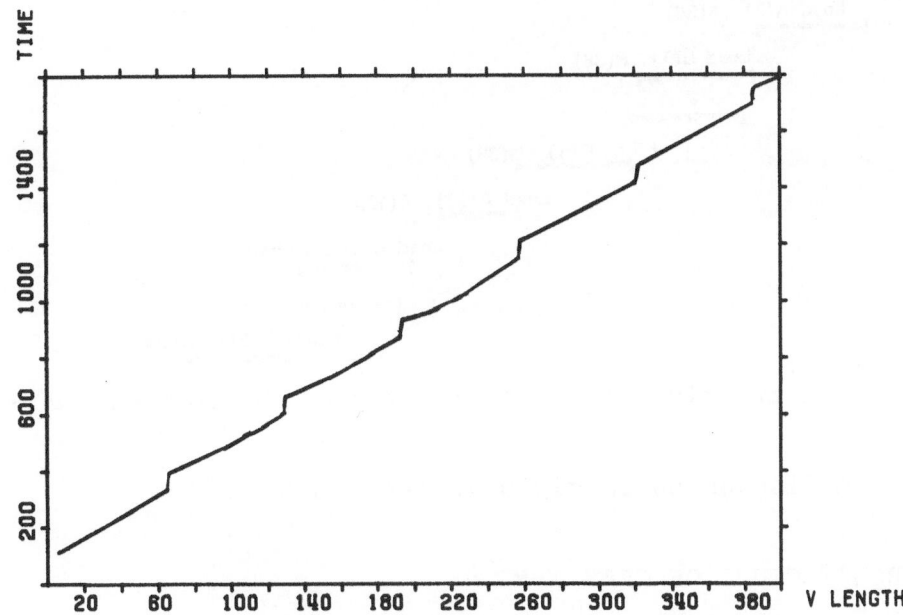

Fig. 2. Timing curve for vector operation.

the sequence of instructions generated by the compiler (CFT 1.11) is similar to this:

```
VL    36        set vector length to 36
V1    A         load 36 elements of A into vector
                register V1
V2    B         load 36 elements of B into vector
                register V2
V3    V1+V2     add V1 and V2 and store result (36
                elements) to V3
C     V3        store result to C (36 elements)
VL    64    ¬
V1    A     │
V2    B     │
V3    V1+V2 ├ process next partial vector of A,B,
            │   and C of length 64
C     V3    ┘
```

The above is a simplified version of the code actually generated by the compiler, because first, no loop is used to process all partial vectors of A and B in the above example and second, all address calculations for indexing A, B, and C have been neglected.

Figure 3 shows the principle execution sequence for the instructions of this example on the CRAY-1. A more

15

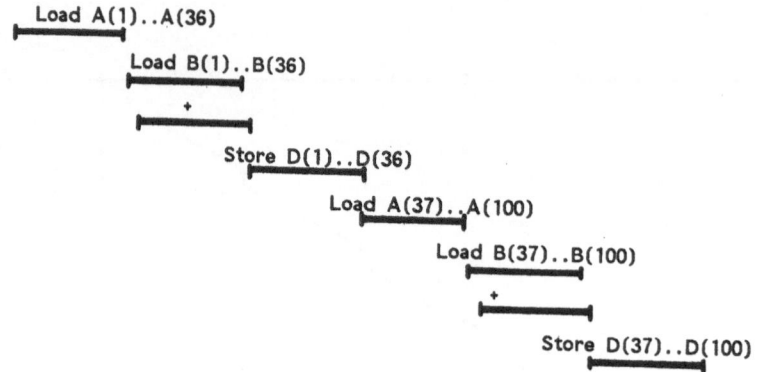

Fig. 3. Instruction sequence for vector addition on CRAY-1.

detailed discussion is given in section 5.

3. ARCHITECTURE OF CRAY X-MP

The most significant difference between CRAY-1 and
CRAY X-MP certainly is the existence of two essentially
independent and identical CPUs in the CRAY X-MP. Both
CPUs of the CRAY X-MP resemble much that of the CRAY-1,
some significant differences, however, should be mentioned. First, the weakness of having only one memory port
in the CRAY-1 has been relieved by the existence of now
four memory ports for each CPU of the CRAY X-MP(figure 4).
By this, one vector store and two vector loads may
execute in parallel in each CPU of the CRAY X-MP. One
port of each CPU is reserved for I/O operations [2][3].

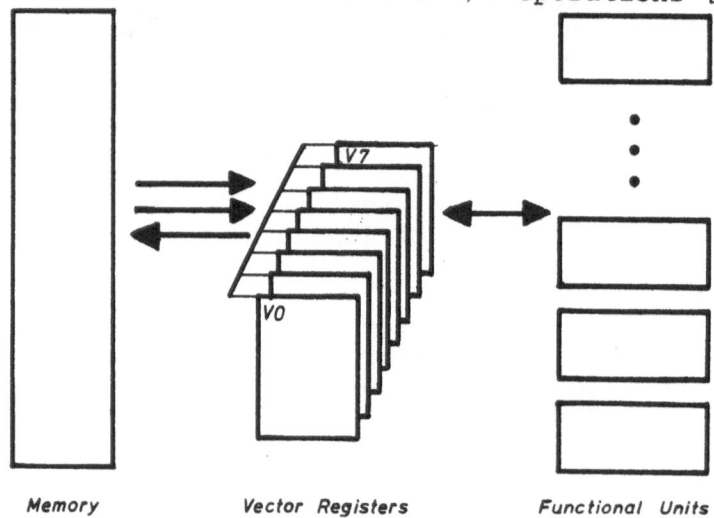

Memory   *Vector Registers*   *Functional Units*

Fig. 4. Data paths for CRAY X-MP CPU.

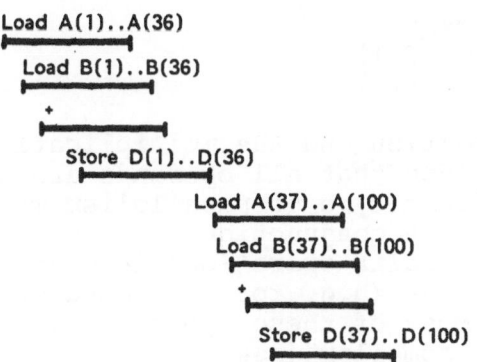

Fig. 5. Instruction sequence for vector addition on CRAY
        X-MP.

The execution sequence for the instructions of the
above example resulting from this difference is shown in
figure 5.

Further differences between CRAY-1 and CRAY X-MP are
the following:

1) faster clock cycle time (9.5 nsec versus 12.5
   nsec),
2) larger instruction buffer,
3) "hardware automatic" chaining.

The first and second of these features are more or
less transparent to the user whereas the third might be
of some interest for the FORTRAN programmer.

4. CHAINING

CRAY CPUs are equipped with 13 independent and
fully segmented functional units used for processing
arithmetic and logical operations. 7 functional units
may be used for vector operations. The most important
ones of these are the Add, Multiply, and Reciprocal
functional unit, which are used for floating-point operations and are shared by scalar and vector operations.

Two ways of parallelism are possible when using
functional units. If distinct operations on distinct
operands are processed as shown in the following example,
several functional units may work at the same time using
distinct sets of registers:

```
     DO 1 I = 1,64
     A(I) = B(I) + C(I)
  1  D(I) = E(I) ⅹ F(I)
```

Here the addition and the multiplication may execute
in parallel provided that all operands are available at
the same time. This type of parallelism may occur for
vector operations in conjunction with scalar operations,
or for vector or scalar operations in conjunction with
address calculations (e.g. for the computation of loop
indices etc.). Most of these types of parallelism are
exploited by the compiler (see section 6) and are transparent to the user. A more sophisticated type of parallelism is described by the notion of "chaining", i.e.
the output of one pipelined functional unit is directly
fed into the next functional unit. This type of parallelism exists for vector operations only.

Figure 6 is a pictorial representation of the
situation caused by the following vector loop, provided
that operands A(I), B(I), and C(I) are available in
vector registers:

```
     DO 1 I = 1,64
     D(I) = A(I) + B(I) ⅹ C(I) .
```

In figure 6, V1 is presumed to hold vector B, V2 to
hold vector C, and V4 to hold vector A. V3 takes up
intermediate result B(I) ⅹ C(I) and V5 the final result
denoted as vector D.

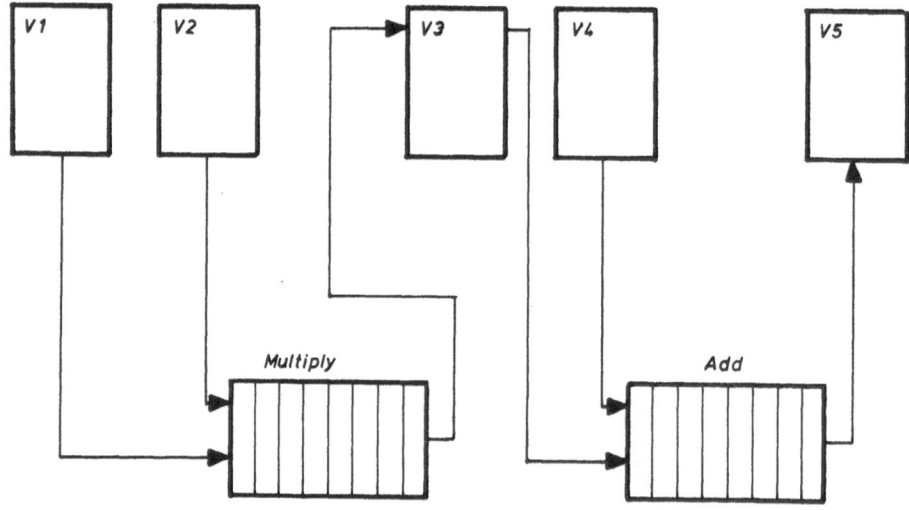

Fig. 6. Chaining.

While there are no major differences between CRAY-1 and CRAY X-MP concerning the parrallelism of independent operations, significant changes concerning chaining have been made in the CRAY X-MP.

On CRAY-1 there is only one pointer for writing vector elements into a vector register or reading them out of it. As a consequence, chaining on CRAY-1 can only be achieved, if the operand (vector element) required for a chained operation is the one just stored into a vector register. For the above example (figure 6) this means that V3 is only able to contribute to the chained operation if the addition begins execution when the first element of intermediate result B(I) $\times$ C(I) arrives at V3. If the addition is issued one clock period too late, chaining is inhibited. The point of time when the addition has to be issued to enable chaining is called "chain slot time". A more detailed description of this situation is given in section 5.

On CRAY X-MP two pointers are used for addressing the elements of vector registers. By this, storing elements into a vector register and reading them out of it may go in parallel for distinct vector elements of the same register. The only restriction is that the "read pointer" may not overtake the "write pointer." By this, chaining on CRAY X-MP is possible at any time. For the example shown in figure 6 this means that chaining occurs independently on whether the addition is already issued when the first intermediate result arrives in V3 or is even not yet ready to issue. In the first case execution of the addition is suspended by hardware until the first vector element of V3 is available, in the second case chaining begins as soon as the addition is issued.

5. EXECUTION SEQUENCE OF FORTRAN KERNELS ON CRAY-1 AND CRAY X-MP

Three Fortran kernels shall be investigated in detail to demonstrate the influence of architecture on vector processing and performance. In order to understand what happens during execution of the FORTRAN codes, it is useful to inspect the sequence of instructions generated by the compiler (CFT 1.11). It is significant that for each example the codes generated for CRAY-1 and CRAY X-MP do not differ from each other. This, however, is not a common property of the CRAY FORTRAN compiler, as some changes have been made in the compiler to take into

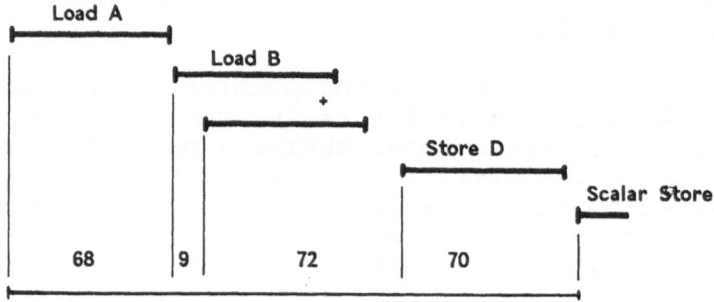

CRAY-1/S:

Load A

Load B

Store D

Scalar Store

68   9   72   70

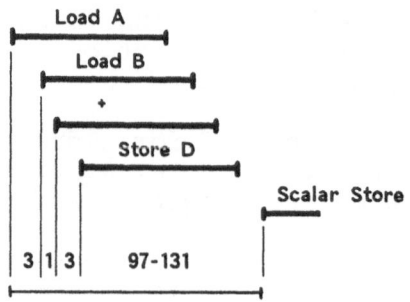

CRAY X-MP:

Load A

Load B

Store D

Scalar Store

3 1 3   97-131

Sum CRAY-1/S:	219 CP
Sum CRAY X-MP:	104 - 138 CP
Quotient:	2.1 - 1.58
x 1.32 :	2.8 - 2.1

Fig. 7. Instruction sequence for vector addition.

account some peculiarities of the CRAY X-MP architecture.

The first example is the same addition of vectors A and B with some vector length N already considered in one of the above examples:

```
    DO 1 I = 1,N
  1  D(I) = A(I) + B(I).
```

The corresponding diagram to this code (figure 7) shows how vector instructions for this vector addition are executed on CRAY-1/S and CRAY X-MP. The diagram reflects one vector loop cycle with a partial vector length of 64.

20

Within this vector loop cycle, first, partial vector A is loaded into one vector register. As there is only one port to memory, the next load for partial vector B has to wait for 68 clock periods (CP) until the memory port is available. Then, the vector load for B can be issued. The next instruction can chain together with the load of vector B. To meet chain slot time the addition is issued 9 CPs after the load for B. The store to memory can start not before the addition is completed, although the memory port is available some time earlier. The reason for this is that the only pointer for filling and reading the CRAY-1/S vector register is occupied by the addition, so that the store operation has to wait. As a consequence, there are three "major cycles" within the vector loop on CRAY-1/S. The end of the last cycle can be determined by testing for the beginning of the next scalar store following the loop.

On CRAY X-MP the situation is quite different. As there are 3 memory ports available for vector loads and stores, loads for A and B can be issued immediately one after the other; there is only a delay of two CPs for some address calculation and one CP for the issue of the load B instruction. The following addition can chain together with both loads. It should be emphasized that the instruction is issued immediately after the load B instruction, regardless of the fact that at this time neither the data of vector A nor of vector B is available. The hardware of the add functional unit itself detects when data is ready. The same is true for the store instruction of result vector D. The instruction is issued immediately and hardware waits until the first result of A plus B is available in the result register. All these wait times sum up in the time given for the duration of the store D instruction; this instruction lasts at least 97 CPs measured from instruction issue. On a fully busy machine the duration can be 131 CPs caused by memory bank conflicts produced by the other CPU of the CRAY X-MP.

Summing up the timings for CRAY-1/S and CRAY X-MP respectively, one obtains 219 CPs for one vector loop cycle on the CRAY-1/S (with a vector length of 64) and from 104 to 138 CPs on the CRAY X-MP depending on whether the XMP is full or empty. The ratio of these values range from 2.1 to 1.6; multiplied with a factor of 1.32 for the faster cycle time on the CRAY X-MP one obtains a factor of 2.8 to 2.1.

CRAY-1/S:

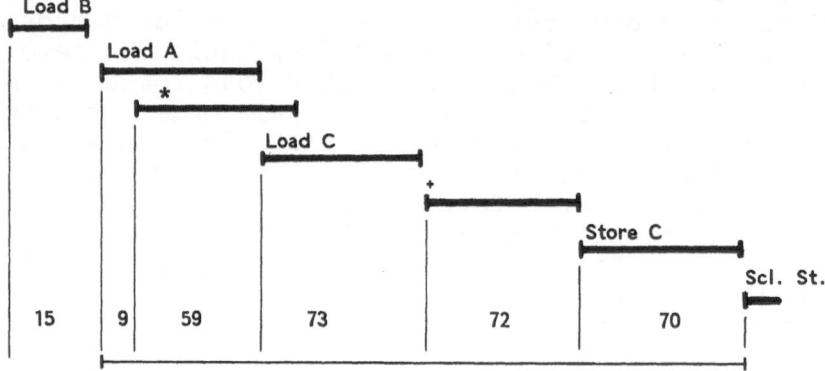

CRAY X-MP:

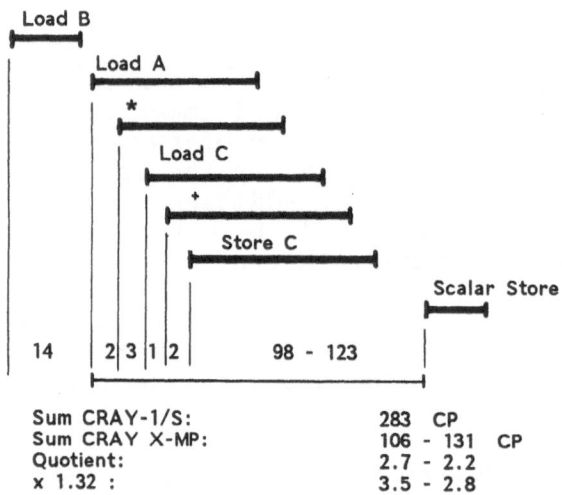

Sum CRAY-1/S:	283 CP
Sum CRAY X-MP:	106 - 131 CP
Quotient:	2.7 - 2.2
x 1.32 :	3.5 - 2.8

Fig. 8. Instruction sequence for triad with scalar.

The next example is a so-called "triad with scalar" as found in the innermost loop of a matrix multiply:

```
      DO 1 I = 1,N
  1   C(I,J) = C(I,J) + A(I,K) ⁒ B(K,J).
```

Here A and C are vectors and B is a scalar for some fixed value of K and J.

Figure 8 again shows the execution sequence of vector instructions for this example. On both machines the load for scalar B is performed outside the vector loop. On the CRAY-1/S the situation is almost the same as for the addition. There is only one difference concerning the addition of C with the product of A and B; namely, there is no chaining for this addition. The

22

reason for this is that the load C operation begins as soon as the port is available. At this time the multiply of A and B is still going on so that chain slot time for the load C instruction is lost when the multiplication has finished.

On the CRAY X-MP the situation is almost the same as in the previous example. There is a full chaining of all operations. So, in this case, there are five vector operations executing in parallel.

Summing up the timings, we obtain 283 CPs for the CRAY-1/S and from 106 to 131 CPs for the CRAY X-MP. Taking into consideration the faster cycle time of the CRAY X-MP this gives a ratio of 3.5 to 2.8 as the speedup of the CRAY X-MP.

CRAY-1/S:

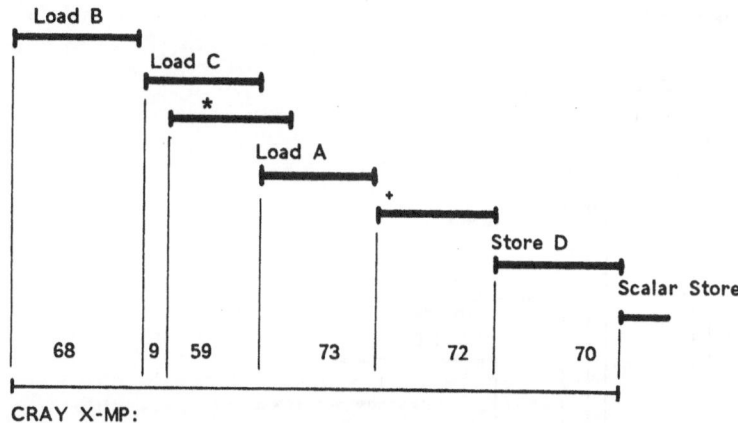

CRAY X-MP:

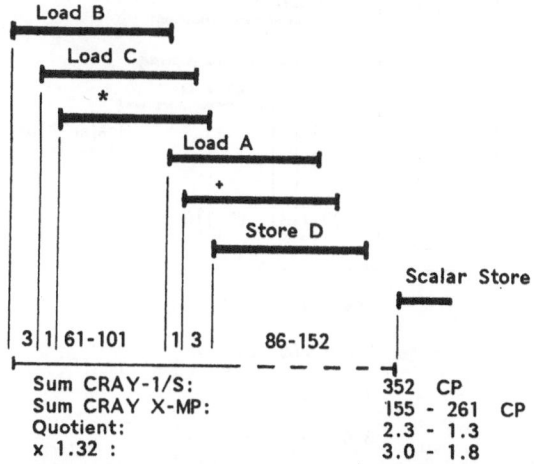

Sum CRAY-1/S:	352	CP
Sum CRAY X-MP:	155 - 261	CP
Quotient:	2.3 - 1.3	
x 1.32 :	3.0 - 1.8	

Fig. 9. Instruction sequence for triad (1).

The last example shows a triad consisting of three vectors A, B, and C:

```
    DO 1 I = 1,N
  1   D(I) = A(I) + B(I) ⋆ C(I) .
```

Figure 9 represents the execution sequence for the instructions of this example.

On CRAY-1/S the situation is exactly the same as for the triad with scalar, except that now the load for B is a vector load and is done within the loop. The rest of the loop is unchanged. On CRAY X-MP there are

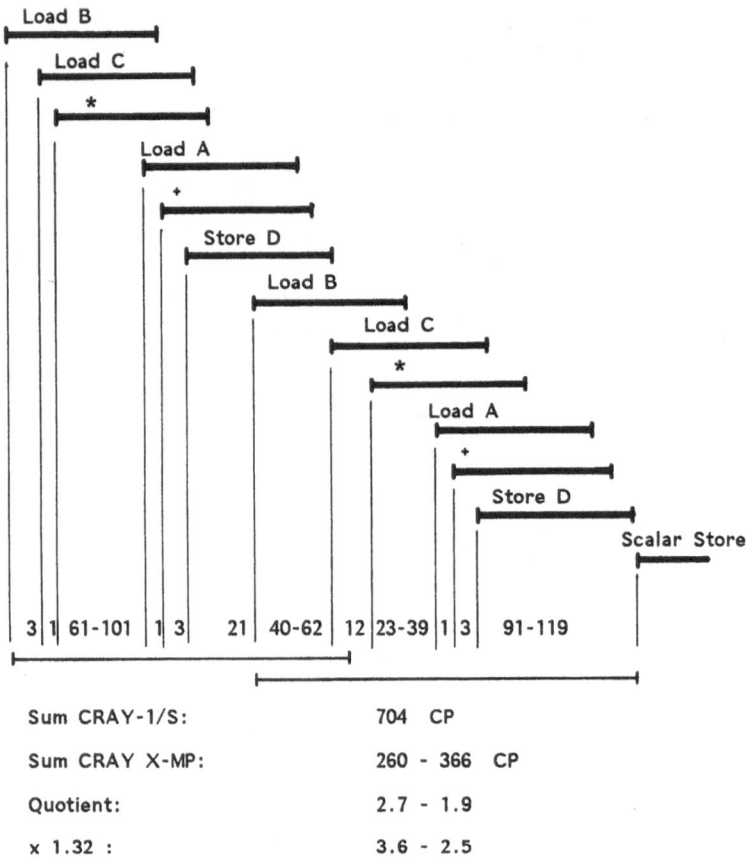

Sum CRAY-1/S:	704	CP
Sum CRAY X-MP:	260 - 366	CP
Quotient:	2.7 - 1.9	
x 1.32 :	3.6 - 2.5	

Fig. 10. Instruction sequence for triad (2).

two major cycles caused by the lack of memory ports. In
the first major cycle, load B, load C, and the multiply
are performed. The load A instruction must wait as there
is nor port available. Then load A, the addition, and
the store D instruction execute in parallel.

Summing up the timing yields a ratio of 3.0 to 1.8
in the comparison between CRAY-1/S and CRAY X-MP. This
however, is not the optimum value that actually can be
observed in the comparison of CRAY-1/S and CRAY X-MP.
Considering not only one vector loop cycle, as done in
the above examples, but two, there is a significant overlapping from the end of the first loop cycle to the beginning of the second cycle.

Figure 10 reflects this fact. As a consequence,
there is a long "chain" of vector operations executing
in parallel beginning with the first loop cycle and
ending with the last one.

At this point it should be mentioned that the above
examples do not reflect typical situations when comparing
CRAY-1/S performance with CRAY X-MP performance. In fact,
a more realistic assessment is obtained by the analysis
of large application programs. Here a value of 1.4 to
1.8 for the ratio of CRAY-1/S CPU times and CRAY X-MP CPU
times is typically found [10].

6. CFT - THE CRAY FORTRAN COMPILER

In principle, the CRAY FORTRAN compiler (CFT)
follows the conception of automatic vectorization. This
includes the detection of vectorizable FORTRAN code as
well as a sophisticated, hardware-dependent rearrangement of the compiler-generated code. The rearrangement
of the generated code is performed by the so-called
"instruction scheduler", a module of the compiler. The
aim is to optimally exploit all types of parallelism:
chaining, parallel execution of independent arithmetic
or logical instructions, and simultaneous execution of
arithmetic and address calculations. Indeed, the latest
compiler version (CFT 1.13) has come up to an excellent
quality concerning this point.

```
    DO 1 I = 1,N
  1  D(I) = A(I) + B(I) :: C(I)
```

for example causes the following instruction sequence
to be generated by the CFT 1.11 compiler (a CFT 1.12
version does not exist) for the CRAY-1:

```
V7    B       Load B into vector register V7
V6    C       Load C into V6
V5    V6×V7   Multiply B and C
V4    A       Load A into V4
V3    V4+V5   Add A and intermediate result V5
D     V3      Store D.
```

This is not optimal as has already been stated in
section 5 (figure 8 and figure 9) because the load for A
starts too early so that chain slot time for the addition
is lost. CFT 1.13 now generates the following code:

```
V7    B       Load B into V7
V6    C       Load C into V6
V5    V7×V6   Multiply B and C
V6    A       Load A into V6
V0    V5+V6   Add A and intermediate result V5
D     V0      Store D.
```

The trick here is that vector A is loaded into
vector register V6 which is reserved as an operand for
the multiplication. As a consequence, the load starts
not before the multiplication is finished. By this,
both operands of the addition are available at chain
slot time and the addition can chain together with the
load for vector A. Figure 11 reflects the instruction
sequence for the improved code.

So, in most cases, programmer intervention will not
be required to make optimal use of the CRAY hardware
concerning chaining, parallelism of independent operations, etc.

As a matter of fact, automatic vectorization has
not come up to such a high quality, although important
improvements have been made during the last years.

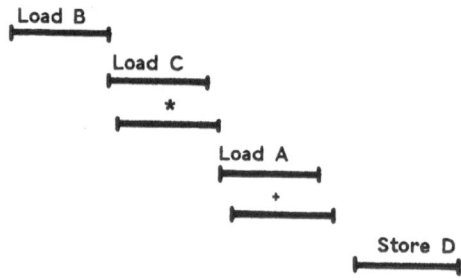

Fig. 11. Improved instruction sequence for triad on CRAY-
         1.

The basic term that serves as a vehicle for expressing parallelism in FORTRAN is the DO loop. Typically, arithmetic operations found in DO loops are transformed to vector instructions, FORTRAN arrays are transformed to vectors. The unrolling of a DO loop into a vector loop operating on partial vectors of length 64 or less is done by CFT and is transparent to the user. A DO loop of length N is transformed into a vector loop with $\lceil N/64 \rceil$ cycles (where $\lceil x \rceil$ is the smallest integer greater than or equal to x). All partial vectors except those processed in the first vector loop cycle have vector lengths equal to 64, those processed in the first cycle have vector lengths of N modula 64 (or 64 if N modula 64 equals zero).

Several conditions must be met to enable CFT to vectorize DO loops. First of all, a linear addressing for all arrays is necessary, as hardware is not prepared to address vector elements that are not regularly spaced in memory. This is the reason why CFT vectorizes inner DO loops but not nested loops.

In principle, the following nested loop could be regarded as one vector operation on vectors A and B, as the assignment of B to A is done for all elements of A and B:

```
    DO 1 I = 1,N
    DO 1 J = 1,M
  1  A(I,J) = B(I,J)
```

The addressing of A and B however, is not linear because A and B are accessed by row but stored by column. For this reason CFT will only vectorize the innermost loop.

Another typical example of nonlinear addressing is represented by the following loop:

```
    DO 1 I = 1,N
  1  A(J(I)) = B(I) ⋇ C(I) .
```

Here B(I) ⋇ C(I) is a typical vector operation with I incremented by one. The indirect addressing of A by index vector J, however, cannot be handled by hardware in vector mode. Yet, CFT 1.13 vectorizes this loop by means of a trick. First, the loop is translated into a vector loop as described above. In each vector loop cycle the term B(I) ⋇ C(I) is evaluated for partial vectors B and C by vector instructions. Now, up to 64

vector elements of intermediate result B(I) ⊁ C(I) have
to be "scattered" to vector A. This is done by a scalar
loop coded within the vector loop. By this, intermediate
result B(I) ⊁ C(I) is scattered to A element by element
in scalar mode. In a way the imbedded scalar loop simu-
lates a vector operation "scatter", not inhibiting vectorization by accepting a partial vector as argument, though
processing it in scalar mode. This type of processing
is called "pseudo vector mode" and is often used in CRAY
library routines.

A last example of irregular addressing shall be
given concerning DO loops containing conditional branch-
ing.

In the following example

```
DO 1 I = 1,N
A(I) = B(I) ⊁ C(I)
IF (A(I).EQ.0) A(I) = B(I)
1   CONTINUE
```

vectorization is possible on CRAY. The conditional IF
statement is handled as follows. For each partial vector
of A a "vector mask" of up to 64 bits is created containing a 1 bit where A is zero, a 0 bit else. Depending on
this vector mask either B(I) is assigned to A(I) or the
old value of A(I) is preserved. This operation is called
"vector merge" (figure 12). It is important to notice

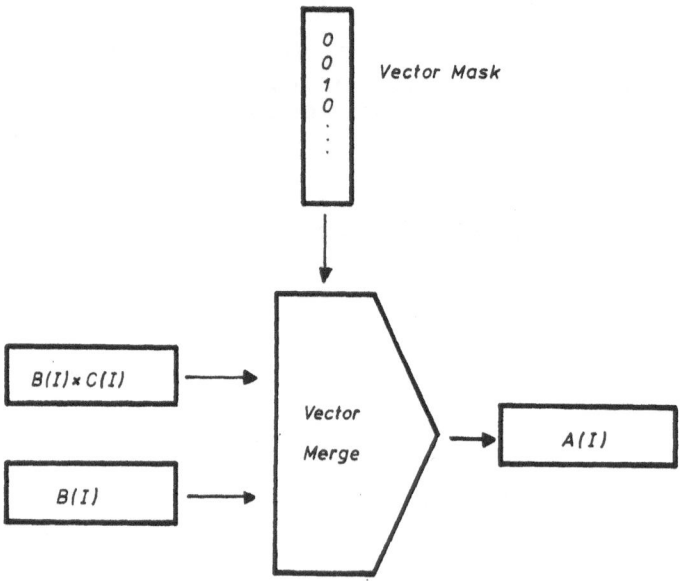

Fig. 12. Vector merge.

that all operations involved in the computation of the conditional assignment are real vector operations. The same principle of operation is possible if the statement on the right-hand side of the IF involves complicated arithmetic. In this case the statement on the right is evaluated first for all elements of a partial vector but now the conditional assignment is performed. This may cause errors if the computation involves divisions and the test was for a zero divide. For this reason the above method of vectorizing IFs cannot always be applied.

7. OPTIMIZING FORTRAN PROGRAMS FOR THE CRAY

A lot of examples might be given where vectorization cannot be done automatically by the compiler. Most occurrences of the following kind require programmer intervention to be handled in vector mode on CRAY:

o   DO loops with calls to subroutines,
o   DO loops with IF and GOTO statements,
o   data dependencies with vector elements,
o   non-linear addressing of vector elements,
o   linear recurrencies,
o   ambiguous subscripts in arrays.

In some cases it can be fairly costly to obtain suitable results. Linear recurrencies as typically found in the solution of tridiagonal systems e.g. may require

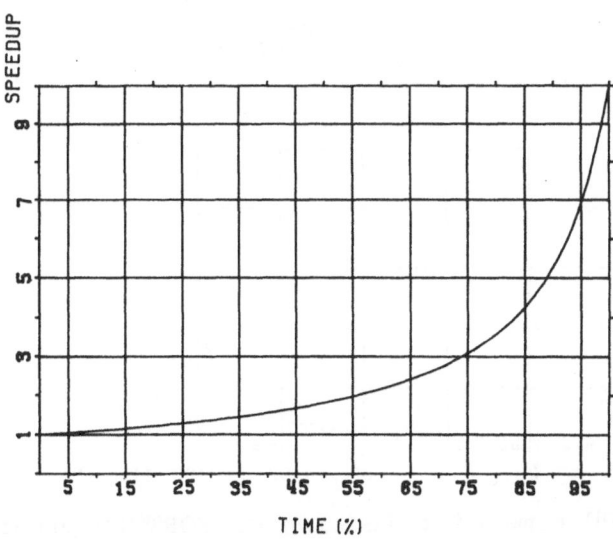

Fig. 13. Speedup through vectorization.

29

intensive modification to be handled adequately on vector computers [4], [5], [6], [7] et al. In other situations automatic vectorization possibly may not lead to optimal results, so that hand optimization of such programs is necessary. Typical actions of this kind are:

o   switching nested loops to enlarge vector lengths,
o   unrolling inner loops,
o   avoid memory bank conflicts,
o   avoid expensive, non-vectorizable types of arith-
         metic if possible (double precision e.g.).

Considering a speedup of factor 10 through vectorization, it is clear that efficient use of vector processors can only be made if a high percentage of the CPU time consumed by a program is spent in vector operations. Figure 13 illustrates this fact. Assuming a factor 10 as the speedup through vectorization, a vectorization of 50% of a given program leads only to a total speedup of factor two, because the non-vectorized part of program will dominate in the run time behaviour. So, at least 80 to 90 percent of the CPU time consumed by a given program have to be transformed into vector operations to achieve a reasonable result.

As a matter of fact, large programs tend to spend a great amount of time in only a small part of the code [8], [9] . Figure 14 shows the distribution of the CPU

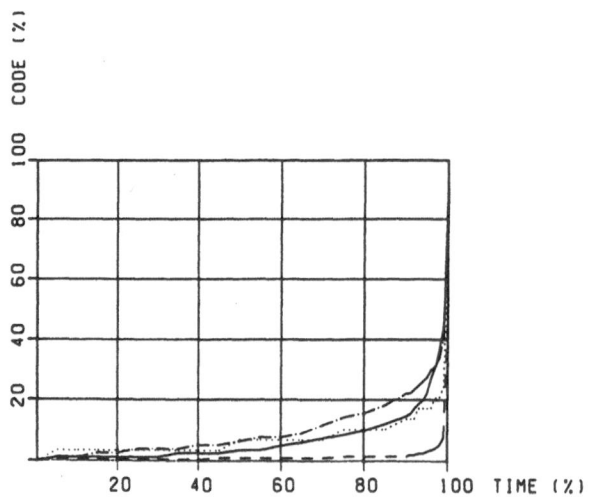

Fig. 14. CPU time distribution of FORTRAN programs.

30

time consumption among the program code for four large FORTRAN programs. Typically 90% of the CPU time is consumed in less than 20% of the code. This can be very helpful for the hand optimization of programs, provided that tools are available that inform the programmer about those time consuming program parts.

8. CONCLUSION

CRAY vector processors allow for several kinds of parallelism when executing FORTRAN programs: parallel processing of vectorized arithmetic operations in the pipelined functional units, parallel execution of independent arithmetic, logical, and address calculations, overlapping of memory accesses, chaining, etc. Most of these types of parallelism are exploited quite well by the CRAY compiler. Yet, when peak performance is desired, a lot of additional hand optimization may be necessary. Tools that supply the programmer with information about the properties of his program may be very useful.

ACKNOWLEDGMENTS

I would like to thank H.W. Homrighausen, W. Oed, and W. Kroj for the contribution of many valuable ideas, hints, and discussions. Particular thanks are addressed to Dr. F. Hossfeld, the director of the "Zentralinstitut für Angewandte Mathematik" for his support and interest in this work.

REFERENCES

[1] W.P. Petersen, Vector FORTRAN for numerical problems on CRAY-1, Communications of the ACM, Vol. 26, No. 11 (1983).
[2] S.S. Chen, Large-scale and high-speed multiprocessor system for scientific applications: CRAY X-MP series, in: "High-Speed Computation," Vol. F7, NATO ASI series, Springer-Verlag Berlin, Heidelberg (1984).
[3] CRAY X-MP Computer Systems, CRAY X-MP Series Mainframe Reference Manual, HR-0032, CRAY Research Inc.
[4] D.E. Heller, D.K. Stevenson, and J.E. Traub, Accelerated iterative methods for the solution of tri-diagonal systems on parallel computers, Journal of the ACM, Vol. 23, No. 4, 636-654 (1976).

[5] P.W. Schwarztrauber, A parallel algorithm for solving general tridiagonal equations, <u>Mathematics of Computation</u>, Vol. 33, No. 145 (1979).

[6] F. Hossfeld, Parallele Algorithmen, <u>in</u>: Informatik Fachberichte, Band 64, Springer Verlag (1983).

[7] W. Oed, and O. Lange, Transforming linear recurrence relations for vector processors, <u>in</u>: Parallel Computing 83, Elsevier Science Publishers-North Holland (1984).

[8] V. Aho and J.D. Ullman,"Principles of Compiler Design", Addison-Wesley Publishing Co. (1978).

[9] U. Detert, and H.W. Homrighausen, Analysis of the dynamic behaviour of FORTRAN programs, <u>in</u>: Proceedings SEAS Spring Meeting 1983, SHARE European Association, Nijmegen (1983).

[10] U. Detert, Performance comparison for CRAY-1/S and CRAY X-MP by means of FORTRAN kernels and user programs, <u>in</u>: Proceedings CRAY User Group Meeting, Paris (1984).

VECTOR PROCESSING TECHNIQUES ON CYBER 205

Karl-Heinz Schlosser

Ruhr-Universität Bochum, Rechenzentrum
Universitätsstr. 150, Geb. NA, D-4630
Bochum, F.R.G.

ABSTRACT

In this lecture we discuss some aspects of algorithm
development on today's vector computers, especially on
the CYBER 205. After a short discussion of the different kinds of parallelism in the architecture of this
machine it is shown, that for the same algorithm the
optimal instruction is a function of the data structure.
Then the weakness of automatic vectorization is shown
and the necessity for semantic vectorization - that means
the development of new algorithms - is pointed out.
Therefore there is a requirement of programming languages
with some parallel features allowing the scientists to
express parallel numerical algorithms in a natural manner.

We will discuss such language kernels and give an
overview of the realization in the new CYBER 205 PASCALV
compiler. Finally we will discuss some numerical algorithms
    (1) iterative linear equation solvers
    (2) adaptive quadrature methods
    (3) numerical optimization with evolution strategies
in respect of semantic vectorization. We will give an
example of the optimal PASCALV implementation of well
suited algorithms and some performance specifications.

1. INTRODUCTION

One of the recent developments in modern sciences consists in the mathematical description of real relationships to get a better understanding of nature. To do this scientists have to define realistic mathematical models of the relationships. But the numerical treatment of these models - especially of dynamic models - normally leads to problems which cannot be solved by general purpose computers. Therefore supercomputers and supercomputer techniques are going to become very important in numerical calculation.

So we can give a first primitive definition: Supercomputers are a class of computers designed for extremely high performance. So there have been supercomputers at any time in the short history of numerical calculation. Looking back one can state that each generation of supercomputers is about five times more powerful than the previous generation and appears 5 - 10 years later. While in the fifties and sixties performance increase of computers was directly proportional to technological progress the situation has changed since the late sixties. Since that time performance increase has been strongly influenced by the progress in computer architecture. This means that the performance of computers depends on the degree of parallelism in the architecture.

In respect of our considerations on vector processing techniques it is important to give a definition, what parallelism means and which kind of parallelism is important for such an architecture. So we define:

Parallelism always refers to an abstraction level of a computer and exists only if several actions on this level can be executed simultaneously.

An abstraction level one can choose e.g. the lowest hardware realization or the instruction level of a computer. For our reflections on parallelism we firstly subdivide computers into several functional units which define their functional structure.

The arrangement of several functional units of the same kind is called horizontal parallelism. A second kind of realization of parallelism is known as pipelining. It can be defined as the technique of decomposing a repeated sequential process into subprocesses, each of which can be executed efficiently on a special dedicated autonomous module operating concurrently with the others

(Ramamoorthy, 1977). The importance of this definition
consists in the partition of the complete action into
several independent subactions with the consequence of
executing these subactions at the same timestep. We
will call this vertical parallelism. One can find these
two structures of parallelism in all modern supercomputers. We will concentrate our discussion on the second
one which can be divided in two subclasses denoting two
special computer architecture levels.

At first the pipelining at the system level which
is typically denoted as instruction pipelining. The
execution of a typical construction consists of

    (1) instruction fetching
    (2) instruction decoding
    (3) operand fetching
    (4) operation performing
    (5) result storing

Using the vertical parallelism principle one can divide
these actions into five segments seg_i, $i=1,..,5$ (see
Fig. 1.1). It is easy to see that seg_i will be free,
if the data of seg_i are stored in the input latches of
seg_{i+1}. For the pipeline cycle time P_τ we get:

$$P_\tau = \max_{i=1..5} P_{\tau i} \qquad (1.1)$$

where $P_{\tau i}$ is the cycle time of segment seg_i. Simplified
the execution time for the above described instruction
without using the vertical parallelism principle is given
by

$$T_B = \sum_{i=1}^{5} P_{\tau i} \qquad (1.2)$$

Using the principle we get

$$T_B' = \sum_{i=1}^{5} P_\tau \qquad (1.3)$$

It can easily be seen that

$$T_B' > T_B \qquad (1.4)$$

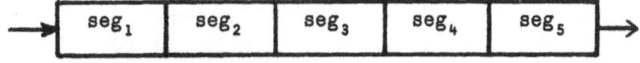

Fig. 1.1. Segments in an instruction pipeline.

Although a single instruction takes a longer execution time using the vertical parallelism principle, one gets a significant performance increase, for after a startup time of $5P_T$ cycles, every P_T cycle the execution of an instruction is finished. This is naturally true assuming that a suitable number of instructions is queued up for execution.

The second level of vertical parallelism consists of pipelining on the subsystem level. One function of the instruction execution process (normally the arithmetic unit) is selected and vertical parallelism is integrated in this function. This principle is often called arithmetic pipelining. It is useful for the execution time of an operation, since P_{T4} is normally big in comparison to the execution time of the other segments.

Let us consider e.g. the segments of a floating point adder which are able to work concurrently:

    (1) sign control
    (2) exponent fitting
    (3) addition of mantisses
    (4) normalization

Combining these two principles leads to a typical architectural realization of parallelism in modern supercomputers, shown in Fig. 1.2. A detailed consideration

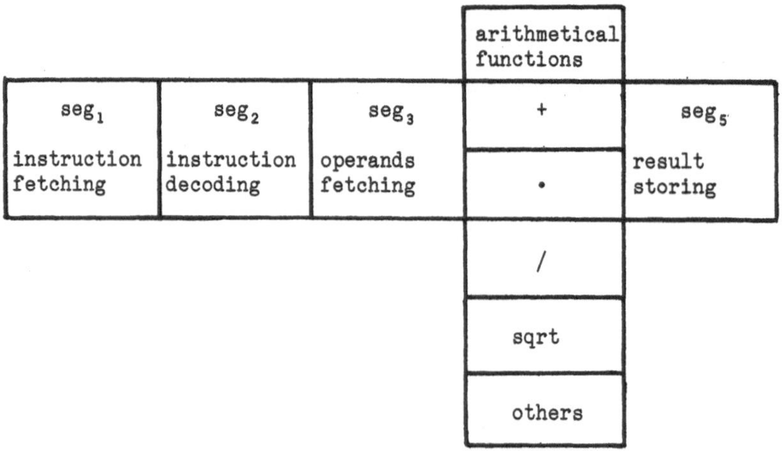

seg_1	seg_2	seg_3	arithmetical functions	seg_5
instruction fetching	instruction decoding	operands fetching	+	result storing
			$\cdot$	
			/	
			sqrt	
			others	

Fig. 1.2. Realization of parallelism in modern super-computers.

36

of this situation involves the following problems:

  (a) A small cycle time in the arithmetic pipelines supposes a suitable cycle time in seg_3 for fetching the operands (respectively result storing in seg_5).
  (b) A suitable number of arithmetic instructions of the same kind has to be queued up for execution.

Normally operations of the linear algebra lead to series of arithmetic instructions which are suitable to condition (b). Therefore supercomputers with such a realization of parallelism are named vector computers.

So all vector computers are characterized by the fact, that the instruction set is extended by a lot of special instructions enclosing the basic operations of linear algebra, the so-called vector instructions. These instructions do not require scalars as operands but vectors.

Under this aspect problem (a) becomes the central problem for computer architectures. The execution of such a vector instruction will only start, if the operands of the instruction - the vectors - are loaded by seg_3. This problem can be solved in different ways:

(w1) Restriction on a short vector length with the following advantages:
  (1) opportunity of storing actual vectors in high speed register sets
  (2) realization of chaining is very easy because of register sets
  (3) short startup times of arithmetic pipelines,
  and disadvantages:
  (1) arising vectors are often broken into small pieces
  (2) each vector instruction with vectors which are not in the register set has to be loaded by a vector load instruction as a preinstruction.

(w2) Abandonment of vector registers with the advantages:
  (1) vector length is in principle unlimited
  (2) necessity of broad data paths to the central memory,
  and disadvantages:
  (1) higher startup times for arithmetic pipelines,
  (2) realization of chaining is more complicated.

Under this aspect the present available supercomputers can be classified as follows:
  Type-w1-computers: CRAY-1, CRAY X-MP, VP100/200, HS81

Type-w2-computers: CYBER 205, VP100/200
All these computers belong to the so-called class VI
computers. We define a class VI computer as follows:

A computer C is element of class VI, if it is able
to generate $(n \cdot 100) \cdot 10^6$, n << 10, n $\in$ N, floating point
results per second of the kind $a := b \underline{op} c$, with
a,b,c $\in$ R" and $\underline{op} \in \{+,-,\cdot,/\}$. R" is the set of floating
point numbers representable on C. This class is some-
times defined as 0(100)-Mflops class.

It is typical for class VI computers that the existing programs written for conventional machines normally
do not reach this performance goal. Before we discuss
our techniques to reach class VI performance on the
CYBER 205, we have to discuss the functional structure
of this machine in respect to the above defined concepts
of parallelism.

Horizontal Parallelism

Taking as a basis for this the maximal extension of
the CYBER 205 one can distinguish four functionally
identical computational units, the so-called pipelines.
All these have the same structure which leads to the
distribution of the elements of the vector instruction
operands:

index of vector element	pipeline
$4 \cdot k + 1$	1
$4 \cdot k + 2$	2
$4 \cdot k + 3$	3
$4 \cdot k$	4

where k $\in$ N $(0 < k < 2^{16})$.

Vertical Parallelism

Both types of this kind of parallelism, i.e. instruction pipelining and arithmetic pipelining are
realized in the CYBER 205. But this can be found in
other supercomputers (e.g. CDC 7600) too. The progress
in the CYBER 205 architecture is founded on the fact
that vertical parallelism on the subsystem level and
horizontal parallelism are able to intensify these
effects. For example let us consider the floating point
addition. We find that a lot of vector elements can be
processed concurrently, since we have five levels of

vertical parallelism in each adding unit (there are four
64-Bit-words). This is also valid for the multiplication.

Besides in the vector processing unit, we also find
vertical parallelism in the so-called scalar processor.
Its adding and multiplying unit are subdivided into five
segments, the shifting unit into three segments. The
other arithmetic units are not segmented.

At the end of our introduction we want to enter into the classification problem and give a classification
of the CYBER 205 to the Erlangen Classification Scheme
(ECS) (Händler, 1975). This is not the only method
(e.g. see Flynn, Feng, Shore, Hockney) but the most
distinguished one and we look upon this scheme as very
useful to show the differences between the architecture
of several class VI computers. In the ECS-scheme the
three logical levels of a computer, the control unit,
the arithmetic unit, and the bandwidth of the arithmetic
unit are described by a triplet where each component
classifies the vertical parallelism as well as the
horizontal parallelism. This gives us:

$$t(computer) = (k \cdot k', d \cdot d', w \cdot w')$$

where k, d, w denote the horizontal parallelism of the
corresponding logical levels and k', d', w' the vertical
one (k-component = control unit, d-component = arithmetic
unit, w-component = bandwidth of the arithmetic units).
Following Händler (1975) we obtain

```
t(Cyber 205) = (
            "control unit" (1,0,0)
            "arithmetic units"
                "scalar unit"
                (+(0,1·2,64·5) "add/multiply unit"
                +(0,1·1,64·3) "shift unit"
                +(0,1·1,64·1) "single-cycle unit"
                +(0,1·1,64·1) "divide unit")
                "vector unit"
                (+(0,4·2,64·5) "add/muliiply unit"
                +(0,2·1,64·2) "shift unit"
                +(0,2·2,64·1) "delay/logical
                                      unit"
                +(0,1·16,64·1) "divide unit")
                "string unit"
                (+(0,1·1,64·5))
```

Considering the fact that the CYBER 205 can process 32-Bit-words too, we can optionally change the number 64 to 32 and double the number of units in the d-component. Thus we have:

$$t(CYBER\ 205) = t(CYBER\ 205\text{-}64\text{-}bit\text{-}version)\ or$$
$$t(CYBER\ 205\text{-}32\text{-}bit\text{-}version).$$

A formal description of the instruction set can be found in Wieczorek (1984).

Now we are able to investigate the consequences concerning algorithms development and programming languages for an architecture like the CYBER 205 machine.

2. SOME REMARKS ON CYBER 205 DATA OPERATORS

Firstly we discuss a taxanomic of the CYBER 205 instructions which was first introduced by Schlosser (1982) and generalized by Peuser (1982) and Wieczorek (1984). The main idea is that every instruction of the CYBER 205 can be interpreted as an operator which operates on an input data set and produces an output data set. These sets may also be empty. We get a useful classification of the instructions by grouping together those instructions which have the same input and output data sets. This leads to the following classification:

    With A := address
          B := bit-vector
          I := integer-number or itemcount
          P := bit-pattern or vector of bit-patterns
          R := register
          S := real number
          V := vector of representable real numbers
          X := vector of representable integer numbers
          β := broadcasting possible
          ↦ := control store possible

we have:

    scalar instructions:
    dyadic floating point   := S × S → S
    monadic floating point   := S → S
    dyadic bit   := P × P → P
    monadic bit   := P → P
    dyadic integer   := I × I → I
    bit compare and branch   := R × R → R

integer compare and branch	$:= (I \times I) \times I \rightarrow (I \times A)$
floating point compare and branch	$:= R \times R \rightarrow A$
bit branch	$:= P \times A$
special branch	$:= A \times (I \times A)$
unconditional branch	$:= A \times A$
integer compare	$:= (I \times I) \times I \rightarrow (I \times P)$
floating point compare	$:= R \times R \rightarrow P$
swap	$:= P \times A \rightarrow P$

vector instructions:

dyadic vector floating point	$:= V_\beta \times V_\beta \overset{Y}{\mapsto} V$
monadic vector floating point	$:= V_\beta \overset{Y}{\mapsto} V$
vector bit pattern	$:= P_\beta \times P_\beta \overset{Y}{\mapsto} P$
special dyadic vector	$:= V_\beta \times V_\beta \mapsto S$
vector assign	$:= P_\beta \mapsto P$
vector maximum	$:= V^\beta \mapsto S$
vector dot product	$:= V \mapsto S$
mash binary compare	$:= P \rightarrow S$
vector index	$:= I_\beta \times I_\beta \mapsto I$
sparse vector	$:= (B \times V_\beta) \times (B \times V_\beta) \rightarrow B \times V$
interval	$:= \emptyset \mapsto V$ or $\emptyset \mapsto X$
search index	$:= V \times V \mapsto X$
scatter and gather	$:= P_\beta \times X \rightarrow P$ or $P_\beta \rightarrow P$
logical bit vector	$:= B^\beta \times B \rightarrow B$
bit compress	$:= B \rightarrow B$
bit merge	$:= B \times B \rightarrow B$
bit mask	$:= \emptyset \rightarrow B$
bit count	$:= B \rightarrow B$
vector compare	$:= V_\beta \times V_\beta \rightarrow B$
single compress	$:= B^\beta \times P \rightarrow P$
vector mask and merge	$:= (B \times P) \times P_\beta \rightarrow P$ or $B \times (P_\beta \times P_\beta \rightarrow P)$
arithmetic compress	$:= V \times V_\beta \rightarrow (V \times B)$
search byte	$:= P \times P \rightarrow I$
move bytes	$:= P \times P \rightarrow P$
select link	$:= V_\beta \times V_\beta \times S \rightarrow V$

For a deeper and more detailed discussion see Wieczorek (1984). We will concentrate our examinations on some selecting operators. The basic data structures of the CYBER 205 are:

(2.1) Vector: A set of data items stored consecutively in memory.
The data items are:
real/integer numbers with item length word = 64 bit

complex numbers with item length double word
= 128 bit
bits with item length bit
short real/integer numbers with item length
halfword = 32 bit.
The cardinality of the set is limited by
65535.

(2.2) A pointer P to a vector - a so called descriptor -
can be defined in PASCAL notation as follows:
const maxint = 2^{47} - 1;
    vecmax = 2^{16} - 1;
type vectorrange = 1 .. vecmax;
    virtualadrrange = 1 .. maxint;
    pointer_to_vector = record of
                length of vector:
                vectorrange
                virtstartadr:
                    virtualadrrange
                end;

Now we can describe some important basic data operators of the CYBER 205:

(a) Control store operator γ:

$$\gamma_B : a \to b \quad \text{where } \gamma_B(a_i) = \begin{cases} b_i & \text{if } B_i = 0 \\ a_i & \text{if } B_i = 1 \end{cases}$$

and a = $(a_1, .. ,a_n)$, b = $(b_1, .. ,b_n)$, $a_i, b_i \in R^*$
are two vectors and BV = $(B_1, .. ,B_n)$, $B_i \in \{0,1\}$
a bitvector. In PASCAL this can be written as:
type realfield = array [1 .. vectorlength] of
                real;
    bitfield = array [1 .. vectorlength] of
                boolean;
...
procedure control_store_operator(var a: real-
                field, b: realfield;
                BV: bitfield);
var index : vectorrange;
begin
  for index := 1 to vectorlength do
    if not BV[index] then a[index]:= b[index]
end;

(b) Vector compress operator σ:
Let a = $(a_1, .. ,a_n)$, b = $(b_1, .. ,b_m)$,
$a_i, b_i \in R^*$ be two vectors with m $\leq$ n, where m
is the number of those elements B_i of bitvector
BV = $(B_1, .. ,B_n)$ which are equal to one.

42

Then σ is defined by:

```
type field_m array [1 .. m] of real;
...
procedure compress (a: realfield; var b: field_m;
                    BV: bitfield);
var index : vectorrange;
    count : vectorrange;
begin
  count := 1;
  for index := 1 to vectorlength do
    if BV[index] then
      begin
        b[count] := a[count];
        count := count + 1
      end
end;
```

(c) Vector mask operator μ:
Let $a = (a_1, \ .. \ ,a_n)$, $b = (b_1, \ .. \ ,b_n)$,
$c = (c_1, \ .. \ , c_m)$, $a_i, b_i, c_i \in R^*$ be three
vectors with $m \leq n$ and $BV = (B_1, \ .. \ ,B_m)$ a
bitvector. Then μ is defined by:

```
type bitfield_m = array [1 .. m] of boolean;
...
procedure mask (a,b: realfield; BV: bitfield_m;
                var c: field_m);
var index : vectorrange;
begin
  for index :+ 1 to m do
    if BV [index] then
      c[index] := a[index]
    else
      c[index] := b[index]
end;
```

(d) Vector merge operator ν:
Under the same assumptions as above ν is defined
by:

```
procedure merge (a,b: realfield; BV: bitfield_m;
                 var c: field_m);
var count_a, count_b, index : vectorrange;
begin
  count_a := 1;
  count_b := 1;
  for index := 1 to m do
    if BV[index] then
```

```
          begin
            c[index] := a[count_a];
            count_a := count_a + 1
          end
        else
          begin
            c[index] := b[count_b];
            count_b := count_b + 1
          end
      end;
```

(e) Gather operator α and the inverse operator
    scatter α^{-1}:
    Let $a = (a_1, \ldots, a_n)$, $b = (b_1, \ldots, b_m)$ be two
    vectors with $a_i, b_i \in R^x$, $n, m \in N$ and
    $I = (u_1, \ldots, u_m)$ an index vector with $u_i \in N$,
    $1 \leq i \leq m$, $u_i \leq n$.

```
    type intfield_m = array [1 .. m] of integer;
    ...
    procedure gather (a: realfield; var b: field_m;
                      I: intfield_m);
    var index : vectorrange
    begin
      for index := 1 to m do
        b[index] := a[I[index]]
    end;

    procedure scatter (var a: realfield; b: field_m;
                       I: intfield_m);
    var index : vectorrange;
    begin
      for index := 1 to m do
        a[I[index]]:= b[index]
    end;
```

We are now able to discuss our main thesis: For the
same algorithms the optimal instruction is a function of
the data structure. To prove this we refer to Bernutat-
Buchmann and Krieger (1982) and consider the very simple
algorithm, given by:

Let $B_q(n,n)$ be a quadratic block of size $n \times n$, which is
part of a $(m \times m)$-matrix. We are interested in elementwise addition of such blocks. This operation shall be
done k times hence we have to consider three independent
variables:
- the blocksize n,
- the supplementary size $d := m - n$,
- the number k of block additions.

There are five different methods to implement this simple algorithm:

- the scalar method,
- the single-vector method,
- the use of control vectors,
- the VXTOV (scatter-gather) techniques,
- the compress method.

Let us now describe these five methods in PASCALV (Ehlich 1984).

1. Scalar Method

```
var A, B, C : array [1 .. m, 1 .. m] of real;
    index_1, index_2 : integer;
    k, n, count : integer;

begin
  for count := 1 to k do
    for index_1 := 1 to n do
      for index_2 := 1 to n do
        C[index_1,index_2] := A[index_1,index_2]
                                + B[index_1,index_2]
end;
```

Using this method we need n^2 scalar additions to solve the problem and therefore - after repeating the algorithm k times - we get:

$$tscalar = n^2 \cdot tadds \cdot k,$$

where tadds is the time for one scalar addition.

2. The Single-Vector Method

In the single-vector method the (n,n) blocks which have to be processed are treated as consisting of n vectors of length n, which are summed up by n vector additions. The resulting PASCAL program has the form:

```
var A, B, C : array [1 .. m, 1 .. m] of real;
    n, k, index, count : integer;

begin
  for count := 1 to k do
    for index := 1 to n do
      C[index,1 .. n]:= A[index,1 .. n]
                           + B index, 1 .. n
end;
```

Theoretically this method takes the time

$$
\begin{aligned}
\text{tsingle} &= k \cdot n \cdot \text{tadd}(n) \\
&= k \cdot n \cdot [\text{sadd} + n/\text{mpls}] \\
&= k \cdot n \cdot [51 + n/2] \\
&= 51\,k \cdot n + k \cdot n^2/2,
\end{aligned}
$$

where tadd(n) denotes the time for one vector addition with vector-length n, sadd the start up time for the vector addition and mpls the machine pipeline size (typically 1, 2 or 4).

3. The Use of Control Vectors

In the next step we try to avoid the n occurrences of the start up time for vector addition. Therefore we take the matrices to be vectors of length u, where

$$
u := n^2 + (n-1)\, d, \quad d := m - n
$$

For such an array we firstly construct a bit vector of length u, where alternating n bits are one and d bits are zero. Then the source vectors are combined by one vector addition. The necessary hardware instructions are firstly a MASKO instruction to produce the bitvector and secondly one vector addition with control vector.

```
type matrix = array [1 .. m, 1 .. m] of real;
var A, B, C : matrix;
    k, n, u : integer;
    ...
procedure contv (var A, B, C: matrix, k, n, u:
                                         integer);
const ADDNV = #82;
      LOD = #7E;
      MASKO = #1D;
      PACK = #7B;
      RTOR = #78;
var bitvec : array [1 .. u] of boolean;
    count : integer;
    aadr, badr, cadr, bitvadr : integer;
    dimension : integer;
    length, rep : real;
begin
  inline (RTOR, A, 0, aadr); ("we need addresses of
                              operands")
  inline (RTOR, B, 0, badr);
  inline (RTOR, C, 0, cadr);
  inline (LOD, bitvec, 0, bitvadr);
```

```
      dimension := m;
      inline (PACK, n, 0, length);
      inline (PACK, dimension, 0, rep);
      inline (MASK0, length, rep, bitvadr);  ("gener.
                                     controlvector")
      for count := 1 to k do
         inline (ADDNV, #00, 0, aadr, 0, badr, bitvadr,
                                          cadr)
   end;
```

This takes the time:

$$
\begin{aligned}
tctl &= k \cdot tadd(u) + tmko(u) \\
     &= k \cdot [sadd + u/mpls] + smko + u/16 \\
     &= k \cdot [51 + u/2] + 45 + u/16 \\
     &= 45 + 51\,k + u/16 + u/2 \cdot k,
\end{aligned}
$$

where $u = n^2 + (n - 1)\,d$ and tmko is the time for one MAS0 instruction with start up time smko.

4. VXTOV (scatter - gather) Technique

A further possibility of solving the problem is firstly to produce two new vectors of length n^2, which consist of the elements of the (n,n) blocks and to sum up these vectors k times with one vector instruction. The resulting vector has to be expanded to matrix form again.

The VXTOV and the compress method are different techniques to generate the new n^2 vectors and to expand the result at the end. In the VXTOV method a gather group instruction is used which takes n groups out of the source matrix each consisting of n elements. After processing the vectors a corresponding scatter group instruction is used for expansion.

```
      procedure scatter_gather(var A, B, C: matrix;
                               k, n: integer);
   const PACK = #7B;
         RTOR = #78;
         VTOVX = #B7;
         VXTOV = #BA;
         LOD = #7E;
   var va, vb, vc : array [1 .. n·n] of real;
       aadr, badr, cadr, vaadr, vbadr, vcadr: integer;
       count, dimension : integer;
       increment : real;
       grlbadr : real; ("group_length_and_base_address")
```

```
begin
  inline (RTOR, A, 0, aadr);
  inline (RTOR, B, 0, badr);
  inline (RTOR, C, 0, cadr);
  inline (LOD, va, 0, vaadr);
  inline (LOD, vb, 0, vbadr);
  inline (LOD, vc, 0, vcadr);
  dimension := m;
  inline (PACK, n, dimension, increment);
  inline (PACK, n, aadr, grlbadr);
  inline (VXTOV, #06, 0, increment, 0, grlbadr, 0,
                                             vaadr);
  inline (PACK, n, badr, grlbadr);
  inline (VXTOV, #06, 0, increment, 0, grlbadr, 0,
                                             vbadr);

  for count := 1 to k do
    vc := va + vb;
  inline (PACK, n, cadr, grlbadr);
  inline (VTOVX, #06, 0, increment, 0, vcadr, 0,
                                             grlbadr)
end;
```

This method costs the time:

$$
\begin{aligned}
tvxtov &= 2\ tgtg(n,n) + tscg(n,n) + k \cdot tadd(n^2) \\
       &= 2\ [sgtg + n\ (48 + n/mpls)] \\
       &\quad + [sscg + n\ (48 + n/mpls)] \\
       &\quad + k\ [sadd + n/mpls] \\
       &= 3\ [26 + n\ (48 + n/2)]\ k\ [51 + n/2] \\
       &= 78 + 144\ n + 3/2\ n + 51\ k + 1/2\ k \cdot n,
\end{aligned}
$$

where tgtg/tscg are the times for gather group and
scatter group instructions with start up times sgtg/sscg.

5. The Compress Method

This method differs from the VXTOV technique by the
way how the operand vectors are produced. Instead of
VXTOV instructions we now use a compress instruction,
which transports the relevant elements from the matrix
into the target vector controlled by a bit vector. The
resulting vector is expanded by a MRGV instruction.
Additionally we have to construct a bit vector with the
help of a MASK0 instruction.

```
procedure compress (var A, B, C: matrix; k, n, u:
                                          integer);
const CPSV = #BC;
      LOD = #7E;
```

```
        MASK0 = #1D;
        MRGV = #BD;
        PACK = #7B;
        RTOR = #78;
    var bitvec : array [1 .. u] of boolean;
        va, vb, vc : array [1 .. n•n] of real;
        aadr, badr, cadr, vaadr, vbadr, vcadr, bitvadr :
                                            integer;

        dimension, count : integer;
        length, rep : real;

    begin
      inline (RTOR, A, 0, aadr);
      inline (RTOR, B, 0, badr);
      inline (RTOR, C, 0, cadr);
      inline (LOD, va, 0, vaadr);
      inline (LOD, vb, 0, vbadr);
      inline (LOD, vc, 0, vcadr);
      inline (LOD, bitvec, 0, bitvadr);
      dimension := m;
      inline (PACK, n, 0, length);
      inline (PACK, dimension, 0, rep);
      inline (MASK0, length, rep, bitvadr);
      inline (CPSV, #00, 0, aadr, 0, 0, bitvadr, vaadr);
      inline (CPSV, #00, 0, badr, 0, 0, bitvadr, vbadr);
      for count := 1 to k do
        vc := va + vb
      inline (MRGV, #01, 0, vcadr, 0, cadr, bitvadr,
                                            cadr)
    end;
```

The theoretical CPU time is:

$$\begin{aligned}
tcpr &= 2\ tcmp(u) + k \cdot tadd(n^2) + tmrg(u) + tmkz(u) \\
&= 2\ [scmp + u/mpls] + k\ \ sadd + n/mpls \\
&\quad + [smrg + u/mpls] + smkz + u/16 \\
&= 2\ [52 + u/2] + k\ [51 + n^2/2] + 58 + u/2 + \\
&\qquad\qquad\qquad\qquad\qquad\qquad\quad 45 + u/16 \\
&= 207 + 25/26\ u + 51\ k + 1/2\ k \cdot n^2.
\end{aligned}$$

Summarizing the results of the different methods we get:

method	CPU time
single vector	$51\ k \cdot n + 1/2\ k \cdot n^2$
control vector	$51\ k + 1/2\ k \cdot u + 45 + 1/16\ u$
VXTOV	$78 + 144\ n + 3/2\ n^2 + 51\ k + 1/2\ k \cdot n^2$
compress	$207 + 25/16\ u + 51\ k + 1/2\ k \cdot n^2$

where $u := n^2 + (n-1)\, d$.

Now we look at the results for several cases:

(c_1) k = 1, n = 100, d = 10
    single vector : 10100
    control vector :   6278
    VXTOV   : 34529
    compress   : 22727

(c_2) k = 1, n = 100, d = 200
    single vector : 10100
    control vector : 11234
    VXTOV   : 34529
    compress   : 36195

(c_3) k = 8, n = 100, d = 10
    single vector : 80800
    control vector : 45099
    VXTOV   : 69886
    compress   : 77786

(c_4) k = 8, n = 100, d = 80
    single vector : 80800
    control vector : 73253
    VXTOV   : 69886
    compress   : 68615

(c_5) k = 8, n = 100, d = 200
    single vector : 80800
    control vector : 144203
    VXTOV   : 69886
    compress   : 84365

In view of these results we have:

case	best method	worst method	speedup of best method against worst method
c_1	control vector	VXTOV	34529/ 6278 = 5.5
c_2	single vector	compress	36195/10100 = 3.5
c_3	control vector	single vector	80800/45099 = 1.7
c_4	compress	single vector	80800/68615 = 1.2
c_5	VXTOV	control vector	144203/69886 = 1.6

It can easily be seen that for each method there is an optimal region (k, n, m) where a given method is the best and another region where it is the worst one. So we can state:

An automatically vectorizing compiler is not able
to generate optimal code since the informations
needed to decide which method of implementing an
algorithm should be chosen, are mostly available
at run time - not at compile time.

Let us now discuss the programming language problems
for supercomputers. Algorithmic programming languages
can be graded in four classes:

C_0 : Languages for sequential machines
C_1 : Compiler for C_0-languages with automatically vector-
    izing possibility
C_2 : C_0-languages with special call features
C_3 : New programming languages for parallel architectures
    or widenings with parallel features

One of the main arguments for C_1-class compilers
refers to the high investments which have been made in
existing software packages. But such an argument has
the consequence that an $0(100)$-MFlops computer at best
is running as an $0(10)$ one. Hence this strategy leads
to the situation that on class VI computers we can only
deal with problems which are solvable on class V computers too.

So the C_3 programming languages should become the
subject of this discussion. At the computer centre of
the Ruhr-University of Bochum a PASCAL compiler for the
CYBER 205 - called PASCALV - has been developed by
Ehlich (1984). This compiler has only few but powerful
extensions to the ISO 7185 standard. These extensions
are:

(a) subranges:
    subrange = expression ".." expression
With this definition a subrange may have variable bounds.
Thus the bound of the index ranges of arrays are no
longer restricted to fixed static values. As a consequence of this all arrays in inner blocks besides the
main program may have variable bounds and by the conform-
ant array scheme this is also valid for the parameters.

(b) index variables:
    index_variable = array_variable
                    "[" index_expression [".." index_
                                           expression]
                    {"," index_expression [".." index_
                                            expression]}
                    "]"

This seems complicated at first glance; the definition
means that a single index expression in the formal definition can be replaced by a subrange with variable
bounds. This simple extension is a powerful tool in
formulating algorithms for modern mathematical methods
in numerical analysis. These mathematical algorithms
often become clearer by writing them in vector notation
and not in coordinates.

example_1: Let a, b, c be three (n×n) matrices and
            c = a ⋇ b
         the matrix product. The normal algorithm for
         computing c is given by a program with three
         nested loops:

```
for i := 1 to n do
   for j := 1 to n do
      begin
        c [i,j]  := 0;
        for k := 1 to n do
            c [i,j]  := c [i,j] + a [i,k] ⋇ b[k,j]
      end;
```

With the inner product operator ip, which is defined in
PASCALV, the inner loop can be written as

    c [i,j] := a [i,1..n] ip b [1..n,j]

or equivalently, the whole algorithms can be written
without any explicit loop as

    c := a ip b.

(c) compatible types:
"Arrays are compatible if they have the same number of
compatible elements." So lower bounds and upper bounds
of arrays are not compared separately, but only the
difference between them.

(d) operations:
The known logical and arithmetic operations for variables
of ordinal type
        and or not + - ⋇ / div mod
are also defined for arrays as elementwise operations.
This definition is recursive. In addition there are
operators for arrays of special type - vectors and
matrices - for the inner product and the matrix product.
The standard functions are treated like operators mapping
the real domain (or a subset) to the real domain. So an
easy extension of the standard functions with array
arguments is achieved. The result of applying standard

functions to an array is an array with elements given by applying the function to the elements of the original array.

With these few extensions in PASCALV the algorithm designer can express the parallelism of the algorithm in a natural way.

For those problems which need the maximal speed of the CYBER 205 there is the opportunity of inserting any machine instruction by a special compiler option similar to special calls in FORTRAN200. In PASCALV this is made possible by using the inline procedure. To give an expression of this feature we consider the following synthetic vector kernel:

```
      DO 100 I,N = 1,N
      A(I) = 0
      DO 100 J = 1,2
      A(I) = A(I) + B(J) ⁒ X(I,J)
  100 CONTINUE
```

This kernel is not autovectorizable. But simple rolling out of the loop leads to:

```
      DO 200 I = 1,N
      A(I) = B(1) ⁒ X(I,1) + B(2) ⁒ X(1,2)
  200 CONTINUE
```

This loop is autovectorizable but needs an additional vector of length N as temporary storage. This problem is solvable in FORTRAN200 by explicitly vector programming. Let AD be the description of vector (array) A, then

```
      AD = B(1) ⁒ X(1,1;N)
      AD = AD + B(2) ⁒ X(1,2;N)
```

is the optimal FORTRAN200 vector kernel because a LINK-instruction is generated. Now we look at the implementation in PASCALV:

```
      var A : array [1 .. n] of real;
          B : array [1 .. 2] of real;
          X : array [1 .. 2,1 .. n] of real;
      begin
        A := B [1] ⁒ X[1,1 .. n] + B [2] ⁒ X [2,1 .. n];
      end
```

Let T_1 := B [1] ⁒ X [1,1 .. n] and T_2 := B [2] ⁒ X [2,1 .. n] be the two terms in the vector kernel. Then there are the following problems:

(1) Since A, X [1,1 .. n], X [2,1 .. n] may be sub-ranges, the compiler has to allocate temporary storage for T_1, T_2 and $T_1 + T_2$.

(2) The compiler is a one-pass-compiler. So it is not able to generate LINK-instructions.

(3) Normally the heap memory is realized on small pages.

To solve these problems there are the following possibilities: (1) and (2) can be solved by using the inline procedure, which generates direct machine instructions. The procedure lpage (lp: boolean) allows the programmer to change the heap allocation strategy. Then the high performance PASCALV-program looks like:

```
type vector = array [1 .. n] of real;
     matrix = array [1 .. 2,1 .. n] of real;
     field = array [1 .. 2] of real;
     pvector = ^vector;
     pmatrix = ^matrix;
     pfield = ^field;

var A : pvector;
    X : pmatrix;
    B : pfield;

procedure kernel (A: pvector; X:pmatrix; B:pfield);
const ADDNV = #82;
      LINK = #56;
      MPYSV = #8B;
      PACK = #7B;
      RTOR = #78;
var b1, b2 : real;
    aadr, x1adr, x2adr : integer;

begin
  inline (RTOR, A^, 0, aadr);
  inline (RTOR, X^ [1], 0, x1adr);
  inline (RTOR, X^ [2], 0, x2adr);
  b1 := B^ [1];
  b2 := B^ [2];
(* vectorkernel *)
  inline (MPYSV, #10, 0, b1, 0, x1adr, 0, aadr);
  inline (LINK, #08, 0, 0);
  inline (MPYSV, #10, 0, b2, 0, x2adr, 0, 0);
  inline (ADDNV, 0, 0, aadr, 0, 0, 0, aadr)
end;
```

This program gives the maximal performance for this algorithm reachable on the CYBER 205, namely about 150 MFlops where $n > 5000$. The PASCALV example has only a performance degree of about 60 MFlops where $n > 2000$.

For a detailed discussion of this kernel and the problem of time measurement see Helmbrecht and Schlosser (1984).

3. NUMERICAL EXAMPLES

In this section we want to give an expression of semantic vectorization of algorithms and show the properties of PASCALV in this connection. One of the most important types of linear systems, the tridiagonal system, arises when solving elliptic problems by grid methods. Lambiotte and Voigt (1975) have studied several well-known methods for solving such systems on the STAR100, a class V computer with a structure similar to the CYBER 205. In the following we want to show a PASCALV implementation and make some performance measurements of the single methods. Consider the linear system of equations

$$A \; x = r$$

with

$$a_{i,j} := \begin{cases} a_i & i = j, & i,j = 1, \,..,n \\ b_i & \text{for} \quad i = j - 1, \; j = 2, \,..,\, n \\ c_i & i = j + 1, \; j = 1, \,..,\, n - 1 \\ 0 & \text{others} \end{cases} \tag{3.1}$$

We discuss the following three methods of solving such a system:

(a) Iterative Gaussian Method of Traub

We make a decomposition of (3.1) in the form

$$A \; x = L \; U \; x = r$$

where

$$L = \begin{pmatrix} 1 & & & \\ q_2 & \cdot & & \\ & \cdot & \cdot & \\ 0 & & \cdot & \\ & & q_n & 1 \end{pmatrix} \qquad U = \begin{pmatrix} u_1 & b_1 & & & 0 \\ & \cdot & \cdot & & \\ & & \cdot & \cdot & \\ & & & \cdot & b_{n-1} \\ 0 & & & & u_n \end{pmatrix}$$

with

$$u_1 = a_1, \, q_i = \frac{c_i}{u_{i-1}}, \; u_i = a_i - \frac{c_i \, b_{i-1}}{u_{i-1}} = a_i - q_i \, b_{i-1}$$

$$i = 2, \ .. \ n. \tag{3.2}$$

First the system

$$L \ Y = r \tag{3.3}$$

is solved. The unknown y_i are given by:

$$y_i = r_i, \ y_i = r_i - q_i \ y_{i-1}, \ i = 2, \ .., \ n \tag{3.4}$$

Then we solve the system

$$U \ x = Y \tag{3.5}$$

and we get

$$x_n = \frac{y_n}{u_n} \ , \ x_i = \frac{(y_i - x_{i+1} \ b_i)}{u_i} \ , \ i = n-1, \ .., \ 1 \tag{3.6}$$

Following Traub (1973) the three basic recurrences (3.1) can be solved by the following iterative method. Define $g_i := c_i \ b_{i-1}$, $i = 2, \ .., \ n$ and $u_i(0) = a_i$, $i = 1, \ .., \ n$ and $u_i(k) = a_1$, $k = 1, \ .., \ M$. Then instead of (3.1) we have:

$$u_i^{(k)} = a_i - \frac{g_i}{u_{i-1}^{(k-1)}} \ , \ k = 1, \ .., \ M, \ i = k+1, \ .., \ n. \tag{3.7}$$

The algorithm will stop after at most M iteration steps, since then the result will be exact. So we can formulate the first part of Traub's algorithm in PASCALV:

```
(AT1) g{[2..n] } := c {[2..n] } ⁑ b{[1..n-1] };
      u := a;
      un1 [1] := u [1];
      i := 1;
      repeat
        i := i + 1;
        un1 [1..n] := a [i..n] - g [i..n] / u [i-1..n-1];
        diff := u - un1;
        diff := diff / un1; (⁑ alignement of exponents ⁑)
        u [i..n] := un1 [i..n]
        error := abs ( max (diff))
      until (error < epsilonu) or (i = n);
```

The algorithm to solve (3.4) is constructed analogously. First we define

$$d_i = \frac{1}{u_i} \ , \ i = 1, \ .., \ n \tag{3.8}$$

to avoid a division. Then we have:

$$y_i^{(k)} = r_i - q_i\, y_{i-1}^{(k-1)} \ , \ k = 1, \ .., \ N, \ i = k+1, .., n$$

where $y_i^{(0)} = r_i$ and $y_i^{(k)} = r_1$. Again the algorithm stops at latest if $N = n$. The PASCALV implementation looks as follows:

```
(AT2) d := 1 / u;
      q := c ⋇ d [1..n-1];
      y := r
      ynl [1] := y [1];
      j := 1;
      repeat
        j := j + 1;
        ynl [j..n] := r [j..n] - q [j..n] ⋇ y [j-1..n-1];
        diff := y - ynl;
        diff := diff / ynl;
        y [j..n] := ynl [j..n];
        error := abs (max (diff))
      until (error < epsilony) or (j = n);
```

At last (3.6) remains to be solved and we define:

$$f_i = y_i\, d_i, \ i = 1, \ .., \ n \ \text{ and } h_i = b_i d_i, \ i = 1, \ .., n-1$$
$$(3.10)$$

Then we have from (3.6):

$$x_i^{(k)} = f_i - x_{i+1}^{(k-1)}\, h_i, \ k = 1, \ ..,P, \ i = 1, \ ..,n-k$$
$$(3.11)$$

where $x_i^{(0)} = f_j$, $i = 1, \ ..,n$ and $x_n^{(k)} = f_n$, $k = 1,..,P$. In PASCALV this can be written as:

```
(AT3) f := y ⋇ d;
      h := b ⋇ d [1..n-1];
      x := f;
      xnl [n] := x [n];
      k := 0;
      repeat
        k := k + 1;
        xnl [1..n-k] := f [1..n-k] - x [2..n-k+1]
                                      ⋇ h [1..n-k];

        diff := x - xnl;
        diff := diff / xnl;
        x [1..n-k] := xnl [1..n-k];
        error := abs ( max(diff))
      until (error < epsilonx) or (k = n-1);
```

This algorithm is convergent for $s + t < 1$, where

$$s = \max_{1 \leq i \leq n-1} |\frac{b_i}{a_i}| \text{ and } t = \max_{1 \leq i \leq n} |\frac{c_i}{a_i}|,$$

(following Traub (1973)).

(b) Vectorized Jacobi Method

The well-known Jacobi method can easily be vectorized and is convergent, if the spectral radius of the iteration matrix is less than one,

$$a_i \, x_i^{(k+1)} = -c_i \, x_{i-1}^{(k)} - b_i \, x_{i+1}^{(k)} + r_i, \quad i = 1,..,n,$$

$$k = 1,..,M \qquad (3.12)$$

For efficiency it is useful firstly to compute $\frac{c_i}{a_i}$, $\frac{b_i}{a_i}$ and $\frac{r_i}{a_i}$, $1 \leq i \leq n$, Lambiotte and Voigt (1975).

The powerful features of PASCALV allow us to give the following canonical implementation:

```
(AJ) x := r / a;
     b := b / a [1..n-1];
     c := c / a [2..n];
     r := r / a;
     i := 0;
     repeat
        i := i + 1; (* iteration counter only *)
        xn1 [1] := -b [1] * x [2] + r [1];
        xn1 [2..n-1] := -c [2..n-1]* x [1..n-2]
                        -b [2..n-1]* x [3..n] + r [2..n-1];
        xn1 [n] := -c [n] * x [n-1] + r [n];
        diff := (x - xn1) / xn1;
        error := abs (max (diff))
     until (error < epsjacobi);
```

(c) SOR-Method

One of the most effective methods for solving such linear systems is the so-called SOR (successive over-relaxation) method. This method has become very important in the last four years in respect to the new multigrid methods for elliptic problems.

We consider the linear system (3.1) and the canonnical partition

$$A = L + D + U \tag{3.13}$$

where

$$L = \begin{matrix} 0 & & & & \\ c_1 & \cdot & & 0 & \\ & \cdot & \cdot & & \\ 0 & \cdot & \cdot & & \\ & & c_n & 0 & \end{matrix} \qquad D = \begin{matrix} a_1 & & & \\ & \cdot & & 0 \\ & & \cdot & \\ 0 & & \cdot & \\ & & & a_n \end{matrix}$$

$$U = \begin{matrix} 0 & b_1 & & & \\ & \cdot & \cdot & & 0 \\ & & \cdot & \cdot & \\ 0 & & \cdot & b_{n-1} \\ & & & 0 & \end{matrix}$$

Then the SOR-iteration algorithm is given by:

$$x^{(n+1)} = D + \omega L)^{-1}\{((1 - \omega)\, D - \omega U)\, x^{(n)} + \omega r\} \tag{3.14}$$

with arbitrary starting vector $x^{(0)}$ and ω $(0,2)$. Writing (3.14) in coordination with $a_{1,1} := a_1$, $a_{1,1+1} := b_1$ and $a_{1-1,1} := c_1$ we get

$$a_{i,i}\, x_i^{(m+1)} = (1 - \omega)\, a_{i,i}\, x_i^{(m)}$$

$$+ \omega\, \{- \sum_{j=1}^{i-1} a_{i,j}\, x_j^{(m+1)} - \sum_{j=i+1}^{n} a_{i,j} x_j^{(m)} + r_i\}$$

$$\tag{3.15}$$

From (3.15) it becomes clear that this algorithm is not vectorizable, because $x_i^{(m+1)}$ depends on the m-th element of the (m+1)-th approximation that have already been calculated. Applying the well-known method of red-black ordering Lambiotte and Voigt (1975) introduced the following modification. We define a permutation matrix P with elements $p_{i,j}$ such that its non-zero elements are given by

$$p_{i,j} := \begin{matrix} 1 & & i \leq \frac{n}{2}\,, & j = 2i - 1 \\ 1 & \text{if} & i > \frac{n}{2}\,, & j = 2i - n \\ 0 & & \text{others} \end{matrix} \tag{3.16}$$

Multiplicating (3.1) with P leads to

$$P\,A\,P^{-1}\,P\,x = P\,r \qquad\qquad (3.17)$$

Writing $y := P\,x$, $z := P\,r$ we get the following transformed system

$$P\,A\,P^{-1} = y\,z \qquad\qquad (3.18)$$

where $P\,A\,P^{-1}$ has the structure:

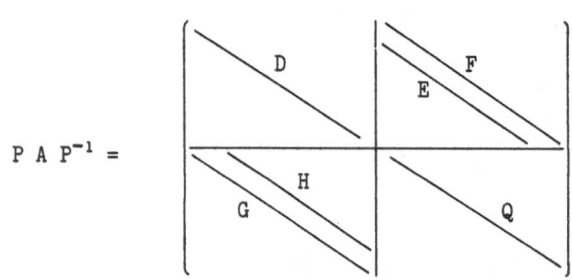

$$P\,A\,P^{-1} =$$

Applying the SOR-method to (3.18) yields:

$$a_{2i-1}\,y_i^{(k+1)} = (1 - \omega)\,a_{2i-1}\,y_i^{(k)}$$

$$= \omega c_{2i-1}\,y_{n/2+i-1}^{(k)} - \omega b_{2i-1}\,y_{n/2+i}^{(k)} + \omega z_i$$

$$(3.20a)$$

$$a_{2i}\,y_{n/2+i}^{(k+1)} = (1 - \omega)\,a_{2i}\,y_{n/2+i}^{(k)}$$

$$= \omega c_{2i}\,y_i^{(k+1)} - \omega b_{2i}\,y_{i+1}^{(k+1)} + \omega z_{n/2+i}$$

$$(3.20b)$$

Now (3.20a) can be computed from known vectors, namely $y^{(k)}$, and then (3.20b) can be evaluated from the known vector $y^{(k)}$ and the first half of $y^{(k+1)}$ obtained in (3.20a). We get the following PASCALV implementation.

```
(ASOR) y := 0;
       i := 1;
       m := n div 2;
       repeat
         yn1 [1..n] := (1 - omega) * d * y [1..m]
                       - omega * e * z [m..n-1]
                       - omega * f * y [m+1..n]
                       + omega * z [1..m] ;
         yn1 [1..m] := yn1 [1..m] / d;
```

```
ynl [m+1..n] := (1 - omega) * q * y [m+1..n ]
              - omega * g * ynl [ 1..m ]
              - omega * h [ 1..m ] * ynl [ 2..m+1]
              + omega * z [m+1..n] ;
ynl [m+1..n] := ynl [m+1..n] / q;
diff := ynl - y;
diff := diff / ynl;
y := ynl
i := i + 1; (* iteration count *)
error := abs ( max (diff))
until (error < epssor);
```

All these methods have been implemented on the CYBER 175 as well as on the CYBER 205. The calculated (theoretical) time for the CYBER 205 has also been added. For a detailed discussion of the results, especially in respect to the SOR-method and time measurement problems look at Helmbrecht and Schlosser (1984).

Table 3.1 and table 3.2 show the computing times for a matrix size of $n = 1000$ and $n = 10000$ where $a_i = r_i = 1$ for $1 \leq i \leq n$ and $b_i = c_{i+1} = s$ for $1 \leq i \leq n-1$. Further definitions are:

$t205$:= time in seconds for CD205
$t175$:= time in seconds for CD175
$tc205$:= time calculated for CD205
ic := number of iterations to get an accuracy of 10^{-12}.

Table 3.1. Computing times for a matrix size n = 1000

s		0.1	0.25	0.49
Traub	ic	(7,13,13)	(11,22,22)	(64,140,143)
	t205	0.0175	0.0283	0.1730
	t175	0.2580	0.4140	2.5900
	tc205	0.0050	0.0078	0.0319
Jacobi	ic	18	41	1410
	t205	0.0113	0.0248	0.8690
	t175	0.1630	0.3470	11.9000
	tc205	0.0020	0.0073	0.1400
SOR	ic	8	13	79
	ω	1.0120	1.0760	1.6700
	t205	0.0096	0.0152	0.0902
	t175	0.1190	0.1910	1.1460
	tc205	0.0036	0.0057	0.0343

Table 3.2. Computing times for a matrix size n = 10000; the results for ic and ω are omitted since they are the same as above.

s		0.1	0.25	0.49
Traub	t205	0.1120	0.1840	1.1350
	t175	2.8700	4.7400	29.3000
	tc205	0.0455	0.0747	0.3506
Jacobi	t205	0.1120	0.1840	5.6600
	t175	1.7600	3.9600	132.0000
	tc205	0.0140	0.0390	1.2800
SOR	t205	0.0526	0.0838	0.4990
	t175	1.2300	1.9600	0.4990
	tc205	0.0349	0.0559	0.3330

For Traub's method a triplet is given denoting the number of iterations for the single steps in the algorithm.

Summarizing we can say:
(1) The factorized SOR-algorithm is the best method for solving such problems.
(2) For all three methods the speed up is about 15 for short vectors and 25 for long vectors.
(3) The maximum speed up for all three methods can only be achieved using the inline procedure, but the gap of performance between PASCALV and PASCALV with inline is not very high (e.g. factor 1.7 at the SOR-method).

In our second example we will consider one of the central problems in numerical analysis, namely the quadrature problem.

Let f : [a,b] → R be integrable. Now we want to compute a best approximation L' of the linear operator L, which is defined by:

$$L : f \int_a^b f(x) \, dx$$

One of the aims in numerical software development consists in constructing an automatical quadrature method. This means, the method operates with the following inputs:

-integration bounds a,b

- method to evaluate the operand f at every point in the interval a,b
- the absolute accuracy $\varepsilon(a)$ or the relative accuracy $\varepsilon(r)$
- an upper bound for the number of function evaluations,

and then produces the following outputs:

- an approximation L' with the accuracy $\varepsilon(a)$ or $\varepsilon(r)$
- an estimation of the real error if possible
- declaring that the problem is not solvable with the given input parameters.

In general such a method consists of

- one or more quadrature formulas
- a method to estimate the real error
- a strategy to control the computation.

It is not the subject of our paper to discuss the problem of error estimation in detail (for this problem see e.g. Schumann (1983) or Davis and Rabinowitz (1975)). We will only describe the general principle which is common to all methods.

By successive application of the basic quadrature formula we obtain - sometimes in subintervals of [a,b] - approximations L_i (i = 1, ...) of L. At each step the approximation L_i is tested in respect to the demanded accuracy until L_i is acceptable or the maximum number of function evaluations is reached.

The basic idea of adaptive methods consists in a dynamic selection of the knots for the function evaluation at each step of the iteration process. One of the main advantages of an adaptive method is the opportunity to integrate "bad" functions (e.g. with singularities). Robinson (1971) and (1976) introduced an interesting method called AGM (Adaptive Gaussian Method), which we will shortly describe:

Let GL(f;I) be the 3-point Gauss-Legendre formula for a subinterval I $\subseteq$ [a,b] and an operand f. The AGM algorithm is given by:

Step 1: L_1 := $[a_1,b_1]$ = [a,b] with subinterval
                                    level J = 0.
        Compute GL(f;I_1).
        Divide I_1 into three subintervals.

$$I_1^{(1)} = [a_1, b_1] \, , \ I_m^{(1)} = [a_m, b_m] \, , \ I_r^{(1)} = [a_r, b_r]$$

with subinterval level J = 1, so that for the three Gaussian knots L_1, M_1, R_1 in the interval I_1 and L_ν, M_ν and R_ν in the intervals $I_\nu^{(1)}$, $\nu = 1$, m, r fulfil the following conditions:

$$M_1 = L_1, \ M_m = M_1, \ M_r = R_1$$

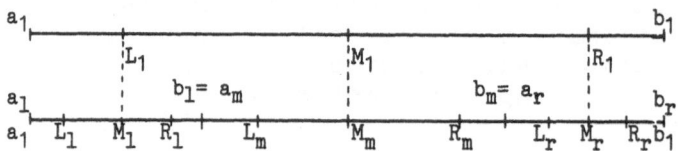

Evaluate $GL(f; I_1^{(1)})$, $GL(f; I_m^{(1)})$, $GL(f; I_r^{(1)})$

$$D(I_1) := \left| \sum_{\nu=1,m,r} GL(f; I_\nu^{(1)}) - GL(f; I_1) \right|.$$

If $D(I_1) \leq \varepsilon(r) \cdot \sum_{\nu=1,m,r} \left| GL(f; I_\nu^{(1)}) \right| =:$

$\alpha(1)$ the algorithm stops and $\Sigma \ GL(f; I^{(1)})$ is an approximation for L. If $D(I_1) > \alpha(1)$ the algorithm goes on to the second step with the interval $I_1^{(1)}$ and the subinterval level J = 1.

Step N: Consider at step N (N = 2,3,...) the interval

$$I_N = I_\nu^{(J)}, \ \nu \in [1,m,r]$$

of subinterval level J and assume that $GL(f; I_\nu^{(J)})$ is known from the previous step. Divide I_N into three subintervals $I^{(J+1)}$, such that for the three Gaussian knots L_N, M_N and R_N in the interval $I_N^{(J)}$ and L_ν, M_ν, R_ν in the intervals $I_\nu^{(J+1)}$, $\nu = 1,m,r$ the following relations are true:

$M_1 = L_N$, $M_m = M_N$, $M_r = R_N$.
Evaluate $GL(f; I_I^{(J+1)})$, $GL(f; I_m^{(J+1)})$,
$GL(f; I_r^{(J+1)})$ and $D(I_N) := \left| \sum_{\nu=1,m,r} \right.$

$\left. GL(f; I_\nu^{(J+1)}) - GL(f; I_N) \right|$.

If $D(I_N) \leq g^{-J} \cdot \alpha(J+1)$ with g := 1.4 then $\alpha(J+1) / \varepsilon(r)$ is a well-usable value over I_N and is added to the approximation L' of L computed in the previous step. In the next

(N+1)st step the right neighbour of I_N is processed.

If $I_N = I_1^{(J)}$ we have to process $I_m^{(J)}$, hence $I_{N+1} = I_m^{(J)}$. If $I_N = I_m^{(J)}$ we have to process $I_r^{(J)}$, hence $I_{N+1} = I_r^{(J)}$. If $I_N = I_r^{(J)}$ there is no right neighbour interval of I_N at the same subinterval level. Therefore it is necessary to reduce the subinterval level by one until an interval

$$I^{(J-J_0)} \text{ or } I^{(J-J_0)}, \quad J_0 \in \{1,2, ..,J-1\}$$

is found with endpoints equal to those of I_N. So

$$I_{N+1} = I_m^{(J-J_0)} \text{ or } I_{N+1} = I_r^{(J-J_0)}.$$

If $D(I_N) > g^{-J}$. $\alpha(J+1)$ then the interval $I_1^{(J+1)}$ of subinterval level J+1 is processed, hence

$$I_{N+1} = I_1^{(J+1)}.$$

If $J + 1 > 39$ (maximum subinterval level), go to the step described above.

The algorithm will stop with an approximation for L if the relative accuracy $\varepsilon(r)$ is reached at a step N_0 on a subinterval I_{N_0} with endpoint b or the upper limit of function evaluations is exceeded.

Robinson (1976) introduced two modifications of the algorithm, which lead to a better convergence:

(1) At each subinterval level an extrapolation step is carried out:

$$\frac{\sum\limits_{\nu=1,m,r} GL(f;I_\nu) - GL(f;I)}{1 - \lambda} \text{ where } \lambda := \frac{47502.\sqrt{15}-183965}{625}$$

(2) The test on accuracy and the sequence of processing the subintervals are modified.

These algorithms have been implemented on a CYBER 205 by Schumann (1983). Schumann got the following results. The CYBER 205 vector processor does not show the opportunity of double precision computing. Therefore four different algorithms had been tested with several functions using 64-bit arithmetic:

VACMV-R : reference algorithm on CYBER 175

VAGMV-AV : reference algorithms on CYBER 205 auto-
vectorized
VAGMV-V : reference algorithm on CYBER 205 scalar
VAGMV-EV : explicite vectorized.

The following table 3.3 shows the results (planned
accuracy: 10^{-6}). The integral was

$$\int_{-1}^{1} \frac{\sin(101 \cdot \arccos(x))}{\sqrt{1 - x^2}} \, dx = 0.\overline{0198} \ .$$

The number of function evaluations was always 2565. The
conclusions are:

(1) The renumication of double precision numbers diminish
    the set of solvable problems especially if "bad"
    operands are given.
(2) Autovectorization of such an algorithm gives cata-
    strophic results (the time is worse than the scalar
    one).
(3) The maximum speed up to a CYBER 175 is less than 5.

So we have to notice that this method is not well
suited for a vector machine of the CYBER 205 type. This
is because the arising vector length is very short.

Our third example shows how vectorization can be
done, if the basic algorithm is not vectorizable. To
point this out we consider the following extreme value
problem.

Given a function $F : R^n \rightarrow R$, find a point $x^* \in R^n$,
for which $F(x^*)$ is minimal (resp. maximal), i.e.:

$$F(x^*) \leq F(x) \text{ resp. } F(x^*) \geq F(x) \text{ for all } x \in R^n.$$

Because max $\{F(x)\} = -\min\{-F(x)\}$ we may restrict our
discussion to the minimization problem without loss of
generality. This means that the optimum has to be found

Table 3.3. The results of Schumann's research

method	value	real accuracy	time (sec)
VAGMV-R	0.019801980713	$2.6 \cdot 10^{-8}$	0.175
VAGMV-V	0.019801980712	$2.6 \cdot 10^{-8}$	0.067736
VAGMV-AV	0.019801980712	$2.6 \cdot 10^{-8}$	0.069563
VAGMV-EV	0.019801980712	$2.6 \cdot 10^{-8}$	0.033172

in a set D with

$$D = \{x \in R^n : G_j(x) \geq 0\}, \ G_j : R^n \to R, \ 1 \leq j \leq m.$$

There are many methods to solve such optimization problems. A detailed comparative description of such algorithms can be found e.g. by Schwefel (1977). Following Bernutat-Buchmann and Krieger (1984) we will introduce a method which has been developed in the field of "bio-engineering". This method has the advantage that no information of F is needed to get a global minimum and that its vectorized version works very efficiently.

In biological terms the evolution strategy algorithms can be described as follows:

Step 0: Given a population of μ individuals. Each individual is characterized by its gene set consisting of n genes.

Step 1: mutation part
The μ parents of the gth generation produce $\lambda \geq \mu$ children. The genes of a child are only slightly different from the corresponding genes of its parents.
The ways in which the children's gene sets are determined (all genes from one and the same parent or recombination of gene sets of different parents) yield different algorithms with different behaviour.

Step 2: selection part
The μ most vital children (depending on the gene set) are chosen. They survive and they will be the parents of the (g+1)st generation. One gets a different algorithm, if not only the children can be selected for the next generation but if also the parents may survive (provided that they are better than their children).

If only children are selected for the next generation the method is called (μ, λ) – otherwise $(\mu+\lambda)$-strategy.

In mathematical notation the algorithms for the (μ, λ)-strategy with recombination can be formulated in the following way:

Step 0: Given
$$e_k^{(0)}, \ 1 \leq k \leq \mu \text{ with } e_k^{(0)} = (e_{k,1}^{(0)}, \ .. \ , e_{k,n}^{(0)})^T \in R^n$$

Furthermore there are m restriction functions

67

$$G_j: R^n \to R, \ 1 \leq j \leq n \text{ with } G_j(e_k^{(0)}) \geq 0,$$
$$1 \leq j \leq n, \ 1 \leq k \leq \mu$$

Set the iteration counter $g := 0$.

Step g: mutation phase
Define

$$r_l^{(g)} := (r_{l,1}^{(g)}, \ .. \ , r_{l,n}^{(g)}), \ 1 \leq l \leq \lambda, \ \lambda \geq \mu \quad (3.21)$$

where $r_{l,i}^{(g)}$ are $(0, s_i^{(g)})$-normally distributed random numbers.
Define

$$p_i^{(g)} := (p_{l,1}^{(g)}, \ .. \ , p_{l,n}^{(g)}) \quad (3.22)$$

where $p_{l,i}^{(g)} = e_{\sigma(e),i}^{(g)}$ and $\sigma: N_\lambda \to N_\mu$ is a random function. The children are determined by

$$c_l^{(g)} := p_l^{(g)} + r_l^{(g)} \quad (3.23)$$

The vectors $p_l^{(g)}$ contain the recombined genes of the parents where the vectors $r_l^{(g)}$ are responsible for the mutation.

Step g: selection phase
Eliminate those $c_l^{(g)}$ with $c_l^{(g)} \notin D$. If λ is chosen too small it might happen that not enough of the generated children fulfil the restrictions. Then the mutation phase has to be repeated. Determine

$$F(c_l^{(g)}) \text{ for all } l, \ 1 \leq l \leq \lambda. \quad (3.24)$$

Order c_l, such that

$$F(c_{l,i}) \leq F(c_{l,j}) \text{ for } i \leq j. \quad (3.25)$$

Select parents for the (g+1)st generation

$$e_k^{(g+1)} := c_{l,k}^{(g)}, \ 1 \leq k \leq u. \quad (3.26)$$

Increment the generation counter $g := g + 1$.
Start the next iteration.

Obviously the algorithm is ideal for computer systems of MIMD-type, where λ or more processors are available, because the operations (3.21) , (3.22) and (3.23) in the mutation phase can be done simultaneously for all l $(1 \leq l \leq \lambda)$. This is also true for the operations (3.25) and (3.26). However this algorithm is typically of an algorithm class, which is also well suited for a super-computer of class VI. To verify this we will formulate the algorithm in terms of vector operations.

To do this all individuals have to be stored into matrices

$$E^{(g)} = (e_{k,i}^{(g)}), \ 1 \leq i \leq n, \ 1 \leq k \leq \mu \text{ for the parents}$$

and

$$C^{(g)} = (c_{1,i}^{(g)}), \ 1 \leq 1 \leq \lambda, \ 1 \leq i \leq n \text{ for the children.}$$

Now we need a random matrix

$$R^{(g)} = (r_{1,i}^{(g)}), \ 1 \leq 1 \leq \lambda, \ 1 \leq i \leq n,$$

a variance vector

$$S^{(g)} = (s_1^{(g)}, \ .. \ , \ s_n^{(g)})^T$$

and an intermediate matrix

$$P^{(g)} = (p_{1,i}^{(g)}, \ 1 \leq 1 \leq \lambda, \ 1 \leq i \leq n.$$

If the data are arranged in this way the following steps can be realized by vector operations:

(a) generation of random numbers $(r_{1,i})$
(b) recombination of parents' genes $p_{1,i}$
(c) generation of children c_1
(d) test of restrictions $G_j(c_1)$
(e) evaluation of the functional values $F(c_j)$
(f) selection of a new generation and assignment
   $e_k^{(g+I)} = c_{1,k}^{(g)}.$

Bernutat-Buchmann and Krieger (1984) had a lot of experience with this algorithm and came to the result that strategies with small μ and large λ seem to be optimal for vector machines.

Summarizing the results of the experiences on the CYBER 205 at the computer centre of the Ruhr-University of Bochum I want to give four theses:

Thesis 1: Autovectorization of serial classical algorithms is normally not a useful way to get class VI performance.

Thesis 2: There is a big requirement for programming languages to express parallelism which is contained in an algorithm in a natural manner.

Thesis 3: The PASCALV compiler for the CYBER 205 is very useful to express parallel algorithms in a natural manner on this machine.

Thesis 4: There are a lot of useful algorithms which are
not useful for the CYBER 205, e.g. adaptive
quadrature methods.

One of the important consequences of these four
theses is the fact, that the class VI computer CYBER 205
should only be used for class VI problems. Those people
who have the impression of a general purpose computer
for number crunching (e.g. a machine to replace a
CD7600) are wrong.

REFERENCES

Bernutat-Buchmann, Ulrike and Krieger, Jost, 1982,
Analysis of elementary vector algorithms, in:
"Conferences on CYBER 200 in Bochum, Proceedings",
Bernutat-Buchmann, U., Ehlich, H., Schlosser K.-H.,
eds., Rechenzentrum, Ruhr-Universität, Bochum,
F.R.G., p. 125-142.
Bernutat-Buchmann, Ulrike and Krieger, Jost, 1984,
Evolution strategies in numerical optimization on
vector computers, in: "Parallel Computing 83,
Proceedings", Feilmeier, M., Joubert, G., Schendel,
U., eds., North-Holland, Amsterdam, p. 99-105.
Ehlich, Hartmut, 1984, "PASCALV - der PASCAL compiler
für den vektorrechner CYBER 205", Bochumer
Schriften zur Parallelen Datenverarbeitung 5,
Ehlich, H., ed., Rechenzentrum, Ruhr-Universität,
Bochum, F.R.G.
Händler, Wolfgang, 1977, The impact of classification
schemes on computer architecture, in: "Int. Conf.
on Parallel Processing 1977", Baer, J.L., ed.,
IEEE Inc., New York, p. 7-15.
Helmbrecht, Detlef and Schlosser, Karl-Heinz, 1984,
Erfahrungen mit PASCALV, in:"Conferences on CYBER
200 in Bochum 1984, Proceedings", to be published.
Lambiotte, Jules Joseph and Voigt, Robert G., 1975,
The solution of tridiagonal linear systems on the
CDC STAR 100 computer, ACM Transactions on Mathe-
matical Software, 1:4, p. 308-329.
Peuser, Michael,1982, Eine systematik der CYBER 205
vektorbefehle, in: "Conferences on CYBER 200 in
Bochum, Proceedings", Bernutat-Buchmann, U., Ehlich,
H., Schlosser, K.-H., ed., Rechenzentrum, Ruhr-
Universität Bochum, F.R.G., p. 107-124.
Ramamoorthy, C.V. and Li, H.F., 1977, Pipeline archi-
tecture, Computing Surveys, 9:1, p. 61-102.

Robinson, I.G., 1971, Adaptive Gaussian integration,
The Australian Computer Journal 3, No. 3, p. 126-
129.
Robinson, I.G., 1976, An algorithm for automatic inte-
gration using the adaptive Gaussian technique,
The Australian Computer Journal 8, No. 3, p. 106-
115.
Schlosser, Karl-Heinz, 1982, Vektoriziering auf sprache-
bene, in: "Conferences on CYBER 200 in Bochum,
Proceedings", Bernutat-Buchmann, U., Ehlich, H.,
Schlosser, K.-H., eds., Rechenzentrum, Ruhr-
Universität Bochum, F.R.G., p. 35-86.
Schumann, Martin, 1983, Vektorizierung adaptiver quadra-
turverfahren, Diplomarbeit, Bochum.
Schwefel, H.-P., 1977, "Numerische Optimierung von
Computer Modellen der Evolutionsstrategie", Birk-
häuser, Basel.
Traub, Joseph F., 1973, Iterative solution of tridiagonal
systems on parallel or vector computers, in:
"Complexity of Sequential and Parallel Numerical
Algorithms", Traub, J.F., ed., Academic Press, New
York, p. 49-82.
Wieczorek, Martin, 1984, "Zur exakten formalen spezifi-
kation der Syntax und Semantik von Machinensprachen",
Bochumer Schriften zur Parallelen Datenverarbeitung
6, Ehlich, H., ed., Rechenzentrum, Ruhr-Universität,
Bochum, F.R.G.

PROGRAMMING DISCIPLINE ON VECTOR COMPUTERS: "VECTORS" AS A DATA-TYPE AND VECTOR ALGORITHMS

Alain Bossavit

Electricité de France
Etudes et Recherches
92141 Clamart, France

INTRODUCTION

Vector machines perform well on vectors. The purpose of this paper is not primarily to explain how and why. If we do so, to some extent, it is only so far as necessary to find a functional characterization of this class of machines. Two concepts seem basic in this respect: extension and regular representation. "Extension", defined below, is a functor by which a scalar arithmetic operation is extended to vectors. "Regularity" means that vectors should be stored at memory locations in arithmetic progression. "Vector operations" are, first the vector extensions of scalar ones, next all operations compatible with the regularity constraint. All of this is explained and justified at full length in the first part of this paper, with many references to the CRAY-1, so this part may be read as a presentation of vector computers, especially the CRAY. But the real point is to define an abstract model of vector computers, rich enough to take their special characteristics into account, yet simple enough not to obfuscate the essential ideas behind vector algorithms.

The "programming discipline" which ensues can be edicted in a few words: forget the idiosyncrasies, program for the abstract machine ("think vectors"), finally work out a Fortran translation. This makes sense only if the programmer already adheres to the kind of "discipline of programming" advocated by Dijkstra (1976), Meyer (1978), to whom I am greatly indebted, Gries (1981), etc.: try to state clearly and rigorously what the problem is first, and go from such specifications down to final programs step by step, actions at one step being similarly speci- fied, to be developed at the next level into more and more detailed programs.

The abstract "vector machine" we describe is, mathe- cally speaking, an algebra. When treating a problem, the ideal approach would be to specify it as some equation written in terms of operators belonging to this algebra. Then, algorithms, or rather algorithmic ideas follow, by simple algebraic manipulations. Of course, we cannot expect this to work for complex programming problems, but it does for large classes of algorithms which happen to be the "elements" from which most complex programs are made: simple linear algebraic algorithms like trian- gularization, solving a triangular system, solving linear recurrences.

By treating these, we shall see that "vector algo- rithms" have no other specificity than the program ob- jects they deal with, namely "VECTORS", in the precise acception we shall indicate. The "algorithmic elements" they are made from are standard ones: iteration, recur- sion, recursion removal and (but this will not appear in the present paper) simple transformation rules. We were especially interested in the analysis, in this sense - that is, finding the underlying algorithmic ele- ments - of well known, but perhaps less well understood, methods for linear recurrence problems like "cyclic reduction" and "recursive doubling", for their general- izations to tridiagonal systems are absolutely basic in numerical analysis. This has been done in some detail. To make up for it, we had to ignore interesting issues, like vectorization of iterative algorithms (relaxation, ICCG, etc.).

The paper is organized as follows:

1. The Vector Machine

    1.1. A look at CRAY-1 performances.
    1.2. Which operations benefit from vector speed and
         why?
    1.3. The "vector machine" concept.

2. Vector Algorithms

    2.1. Linear recurrences.
    2.2. Linear algebra.
    2.3. The "total reduction" problem (recurrences in a
         monoid).

A shorter version was presented at the IFIP Working Group 2.5, at Söderkoping (Sweden) in the fall of 1983. The discussion on recursive doubling and cyclic reduction is new.

THE VECTOR MACHINE

"Vector computers" (CRAY-1, CYBER 205, etc.) perform better on <u>arrays</u>, or "vectors" of data, than on the isolated data. We shall see this on a typical example, then introduce the relevant concepts of "pipe-line" and "vector machine."

1.1. A Look at CRAY-1 Performances

Consider the Fortran statement

$$A = B + C \tag{1}$$

where A, B, C are entities of REAL type, and, concurrently, the sequence

```
      REAL A(...), B(...), C(...)
C     -- N IS A POSITIVE INTEGER
      DO 1 I = 1, N                              (2)
1         A(I) = B(I) + C(I)
```

and let us measure their execution time on a vector computer, the unit being the "clock-period" (CP), which is 12.5 nanoseconds on CRAY. Let us denote by S the timing result for (1), in CP, approximately 30 on CRAY-1. As for (2), the approximate formula

$$t(N) = I + N V \tag{3}$$

is roughly consistent with the timing results (cf. Fig. 1), with V = 3 and I around 70 ("I", "V" and "S" stand for Initialization, Vector, Scalar (time) respectively). If the machine were treating (2) just as N additions like (1), i.e. repeated N times the same actions on different data, the measured time would be N S.

This simple example is enough to understand how vector computers are functionally different from conventional ones. In many cases, when the same action (like the addition above) has to be applied to a series of N sets of data with identical structure, N large enough, one can take advantage of another execution mode ("vector"

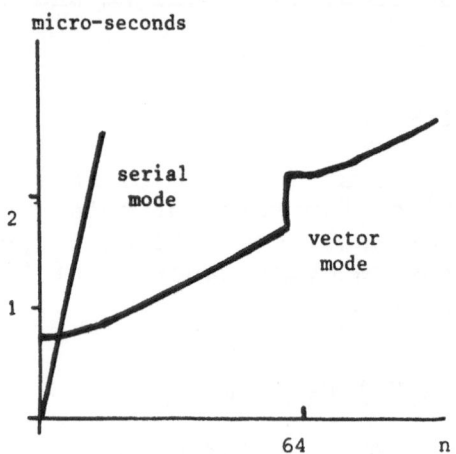

Fig. 1. Typical timing curve on CRAY-1.

OPERATION	LOOP BODY	S	I	V	R	I/V
assign	$Y(I) = X(ADR(I))$	41	0	41		
assign	$X(I) = X1$	22	83	2		
recurrence	$X(I) = A*X(I-1)+Y(I)$		63	37	0	
scalar + vector	$Y(I) = A + X(I)$		69	2		34
vector/scal.	$Y(I) = X(I)/A$	73	111	2	19	55
vector divide	$X(I) = Y(I)/Z(I)$	73	138	6	61	23
add	$A(I) = B(I) + C(I)$	45	103	3	26	34
multiply	$A(I) = B(I) * C(I)$		104	3	26	34
saxpy	$X(I) = A*Y(I) + Z(I)$	44	106	4	26	26
saxpy	$X(I) = A*Y(I) + (Z(I))$		115	3	35	38
Vabs	$X(I) = ABS(Y(I))$		92	2	18	46
Vsin	$X(I) = SIN(Y(I))$	347	408	28	275	15
	$Z(I) = X(I)*(X(I)+Y(I))$		114	4	35	29
	$Z(I) = X(I)*X(I) +$					
	$\qquad\qquad X(I)*Y(I)$		114	3	35	38
sum	$S = S + X(I)$		339	3		113
dot product	$S = S + X(I) \, Y(I)$		319	5		64
	$X(I) = X(I - 1)$		0	22		
	$X(I) = X(I-1) + Y(I)$		0	32		

Fig. 2. A few timing results on CRAY-1. See text for
S, I, V, R. The unit is the Clock Period (CP),
i.e. 12.5 nanoseconds.

mode), much faster than the mere repetition of the action on the N data sets ("sequential" mode). When such a situation occurs, a natural formalization consists in thinking of the arrays of data as "vectors" of the same length, the operation being performed on the set of components with the same index, for all values of this. In short: vector computers are faster on <u>vectors of data</u> than on <u>isolated</u> (or irregularly ordered) data, provided the vectors are long enough.

Quantities I, V and S obviously depend on the action (cf. Fig. 2). But the ratios S/V and I/V happen to be roughly constant, and are characteristic parameters for a given machine. On CRAY-1, S/V is around 7 to 10, and most vector machines have comparable S/V ratios. On the opposite, I/V has a much wider spectrum, ranging from 20 to 30 on CRAY to a few hundred on CYBER 205, and a few thousand on ICL's Data Array Processor (DAP). This ratio is called "$n_{1/2}$"by Hockney (1981), for "half-performance length", the vector length for which execution speed is just half the maximal (asymptotic) speed.

Hockney has made $n_{1/2}$ the corner-stone of a classification of vector computers (the bigger it is, the more parallel the machine) and the basis of a method of a priori evaluation of algorithms. One should be warned against a possible confusion: $n_{1/2}$ is not the "break-even" point, vector length for which serial and vector mode take the same time. This length is close to S/V, the other dimension-less parameter.

Let us now have a closer look at Fig. 2. Each row describes a "vector operation", in the usual mathematical acceptation of the word. For instance, the row labelled addition reports measurements done on code (2) above, and "loop-body" refers in this case to statement 1 in (2). Similarly, the row which has

$$Y(I) = X(ADR(I)) \tag{4}$$

as loop-body corresponds to the following program section

```
      INTEGER ADR(...)
      .....
      DO 1 I = 1, N                              (5)
          J = ADR(I)
  1       Y(I) = X(J)
```

assuming that array ADR contains some permutation of the integer segment [1..N]. In a slightly higher level notation, (5) could be transcribed as

$$y \to perm(adr, x) \tag{6}$$

("assign to variable y the value of the expression on the right"). One says that this operation is of the following "type":

$$INTEGER\_VECTOR \times VECTOR \to VECTOR \tag{7}$$

A mathematician would claim this is a vector operation, since it amounts to multiplying vector x by some permutation matrix. But obviously it is performed at serial speed only.

The row $X(I) = X(I-1)$ is in the same category, but will take more of our time. Let us introduce the notation "τ". If $x = \{x_1, x_2, \ldots, x_n\}$ is a vector of n components, then

$$\tau x = \{0, x_1, x_2, \ldots, x_n\}, \tag{8}$$

a (n+1)-vector obtained by right-shifting x. The loop

```
C   -- X(1) HAS ALREADY BEEN ASSIGNED SOME VALUE AT THIS
          POINT
    DO 1 I = 2, N                                        (9)
1        X(I) = X(I-1)
```

does not translate

$$x \leftarrow \tau x \tag{10}$$

as all Fortran programmers know (having forgotten how natural the converse looked to them, before they were indoctrinated). What (9) does, by the way, is

$$x \leftarrow broadcast(x_1, n). \tag{11}$$

Here, broadcast: REAL × INTEGER → VECTOR consists in building a n-vector whose components are all equal to the content of variable X(1), namely x_1.

Compare now with the row where precisely this assignment $X(I) = X1$ appears. As we see, (9) is very inefficient coding for (11). This is a simple example (though a bit artificial) of what happens when programming on vector computers: a drop in efficiency by a factor of 10 due to awkward coding.

Assignment (10) will give us another example.   It can be translated this way (TOLD and TNEW are two auxiliary variables):

```
     TOLD = 0
     DO 1 I = 1, N+1
          TNEW = X(I)                              (12)
          X(I) = TOLD
          TOLD = TNEW
   1  CONTINUE
```

or that way (now one makes use of a whole array T of temporaries)

```
     DO 1 I = 1, N
   1      T(I) = X(I)
     X(1) = 0                                      (13)
     DO 2 I = 2, N+1
   2      X(I) = T(I-1)
```

These two sequences have the same outcome in any Fortran environment. But on a vector computer, (13) is much faster, as Fig. 2 testifies. The trouble is that one would rather spontaneously use (12), if only to save memory space. Yet (13) is more natural, in some sense, being the direct translation to Fortran of

$$t \leftarrow x \; ; \quad x \leftarrow \tau t \tag{14}$$

where t and x are vectors.

This suggests the potential interest of a programming discipline by which one would strive to identify vectors and to consciously manipulate these through vector operations. Of course, since the corresponding instructions do not (yet) exist in Fortran, some coding in terms of lower-level instructions (ordinary scalar arithmetic) would follow. Such a discipline, which is already desirable for the sake of readability, clarity, resistance to error, ease of maintenance, etc., does not contradict efficiency.

Well, this is what we are after. But we are not through with Fig. 2 yet. A group of operations have been isolated which, though mathematically of "vector" type, execute at serial speed, either because of erroneous coding, or, e.g. in the case of (4), for reasons which have still to be made explicit. Among the other operations (those with a small V), there is a minority which exhibits a high initialization time: dot-product, sum of components. Thus we can distinguish three kinds of operations:

- the "vector" ones, stricto sensu (I and V small),
- the "pseudo-vector" ones (V small, I large),
- the "serial" ones (V = S and I negligible).

Let us now outline our approach. Only the operations of the <u>first group</u> will be accepted as bona fide "vector operations". We shall see they form a family large enough, complete enough, so that one can program in terms of these operations, in priority if not exclusively. The idea is thus, instead of programming "towards" what can be called a "Fortran machine", as we usually do, to code consciously towards some abstract "vector machine", characterized by this family of operations. Operations of the second group and a fortiori those of the third, will not be invoked before having tried all the possibilities of the vector machine. (No quotation marks from now on, but let's not forget the vector machine is a virtual one, an abstract model for real machines.) A phase of translation to Fortran will of course occur at the end. By this "discipline of programming", one will automatically achieve fast execution, in addition to other advantages inherent in such an approach.

The motto will thus be "think vectors".

To obey it consistently, we must first understand what distinguishes "true" vector operations. This will also explain how vector machines can be that fast.

1.2. Which Operations benefit from Vector Speed, and why

1.2.1. Extensions. This concept is borrowed from the programming language APL. Let $\oplus$: SCALAR $\times$ SCALAR $\rightarrow$ SCALAR be an operation. (The scalar type can be virtually anything, but we have in mind types like REAL, INTEGER, and $\oplus$ can be add, multiply, etc. or else LOGICAL, $\oplus$ being one of the relationals .AND., .OR., etc.) (The number of operands, here 2 for definiteness can be different). Let u and v be two vectors, i.e. two arrays of scalars of common length n. The vector extension $V\oplus$ of $\oplus$ is defined as the vector w = $V\oplus(u,v)$ of length n whose components are $w_i = \oplus(u_i, v_i)$.

Example: the extension of sin associates with vector u the vector v = Vsin(u) of components sin u_i. The "V" discriminant will be omitted without any danger of confusion in most cases, as in z ← (u+v)∗w (where u, v, w and z are vectors of the same length). Remark that this expression can be understood either as the extension of the scalar operation a ← (b+c)∗d, or as the <u>composition</u> of V+ and V∗.

On a serial machine, the straightforward way to implement an extension is by a loop:

<u>for</u> i := 1 <u>to</u> n <u>do</u>
$$\oplus(u_i, v_i, \ldots)$$

and the parallelism between the n actions is left unexploited.

This "extension" concept is helpful in our exploration of Fig. 2: Vector operations (stricto sensu) are precisely vector extensions of conventional arithmetic or logical operators, of standard library functions, and of operations obtained by composition from these. Now we have a simple and precise characterization of vector computers: those machines which are relatively <u>more efficient on extensions of scalar operations</u> than on these taken piece by piece.

<u>Remark</u>. Another APL concept, namely "reduction", is ready-made to characterize operations of the second group. If ⊕ is a binary operation with a neutral element e, its "vector reduction" is defined by the outcome of the following program:

/⊕ (<u>in</u> u : VECTOR, <u>out</u> S : SCALAR {n=length(u)})
  s ← e ;
  <u>for</u> i := 1 <u>to</u> n <u>do</u>   (15)
    $$s \leftarrow s \oplus u_i$$

For instance, the sum of the components of a vector is the vector reduction of addition. So we have, as an observational fact, this: On vector machines, vector reductions (and operations obtained by composition including at least one of these) <u>suffer from high start-up times</u>, though, on the other hand, their <u>asymptotic speeds</u> are of the <u>same magnitude</u> as those of extensions. Why this is so - though (15) would suggest scalar speed only - will not be explained until we see the linear recursion problem.

<u>1.2.2. Arrays of processors and pipe-lines</u>. How to build a vector machine? By the application, possibly simultaneously, of two diametrically opposed principles:

- the <u>multiplication</u> of functional units,
- the <u>segmentation</u> of functional units.

"Functional units" in this context, means the hardware parts which perform scalar operations. For each operation like +, ×, etc., there is a specialized unit (even if this specialization is achieved by micro-programming, we shall pretend the units are separate).

The "multiplication" idea is clear enough: n identical units, or "processors" will do on n-vectors what one of them would do on scalars in the same amount of time. "Pipe-lining", on the other hand, is just the application to computing of the assembly line concept (see Ramamoorthy and Li, 1977). We shall describe the two procedures in a rather more abstract way than usual.

The problem is thus to conceive an extension as a combination of lower-level operations, corresponding to existing hardware. Few scalar actions $\oplus$ are actually elementary: a simple addition of two floating-point numbers, for instance, is made of a series of lower-level actions: fetch operands, compare exponents, align mantissas, add, etc. If there are m such steps, the action $\oplus$, when done on components indexed by i, consists in m successive elementary actions A_i^1, A_i^2, ..., A_i^m. So the extension of $\oplus$ consists of nm partial tasks A_i^j, whose order of succession is not indifferent. Figure 3 shows the "dependence graph" of these tasks: each corresponds to a vertex, and an arrow from A to A' means that "A' depends on A" for its execution, for instance because results produced by A are data for A'. Such a diagram evokes a <u>scheduling</u> problem. Suppose for simplicity all elementary tasks require the same execution time, which will serve as time unit. The problem is to assign each task a starting time in such a way that dependencies are respected and no conflict in accessing resources occurs, while of course minimizing the overall execution time.

"Resources" here means the hardware necessary to complete a task. On a classical computer, such resources are scarce (typically, only one adder is available), so the execution order is uniquely determined, up to a permutation on i-indices: this order is shown in Fig. 3b. If resources abound (multiplication of functional units), the order can be that of Fig. 3c, where tasks with the same j-index start simultaneously. But of course, there are a lot of other possible schedulings, for instance Fig.

$$A_1^1 \rightarrow A_1^2 \rightarrow A_1^3 \rightarrow A_1^4$$

$$A_2^1 \rightarrow A_2^2 \rightarrow A_2^3 \rightarrow A_2^4$$

$$\ldots \rightarrow \quad \ldots$$

$$\ldots \rightarrow \quad \ldots$$

$$A_n^1 \rightarrow A_n^2 \rightarrow A_n^3 \rightarrow A_n^4$$

(a)

Dependence graph of the A_i^js.

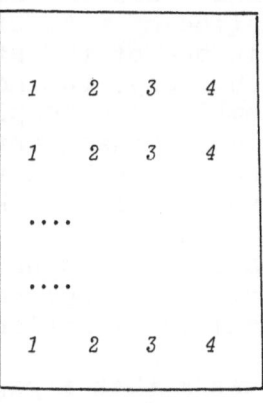

(b)

Serial scheduling of
the tasks.

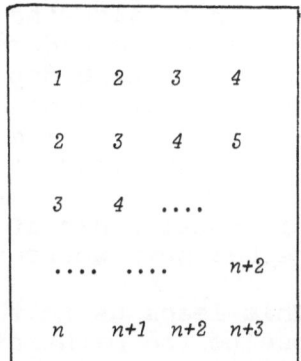

(c)

Scheduling if n parallel
processors are available.

(d)

"Pipeline" scheduling.

Fig. 3. Dependence graph for the vector extension of a
scalar operation which executes in m steps,
m = 4, and three compatible schedulings.

3d. This particular one saves resources (no more than m active tasks at a time), as opposed to solution c, and time (n + m - 1 units, instead of nm for the serial solution).

This "oblique" scheduling can be implemented by the association of m functional subunits, each one specializing in one of the steps A^j of operation $\oplus$. They work simultaneously, but not on the same index i, just as in an assembly line people work on different cars at the same time. Consequently, some transfer of data must exist between sub-units: number 1 passes on to number 2 the result of its work on components i, while 2 gives to 3 what it did on i-1, etc. The sub-units should thus be physically contiguous. Like segments of an insect body, they form, put together, a functional unit able to perform the totality of action $\oplus$. This is the pipeline concept.

Remark that a pipeline unit, though more efficient on vectors of data, can also perform scalar operations. In some machine architectures, scalar and vector operations are done by the same functional units. In others, Cray's for instance, there are distinct scalar and vector units. Beware also that in our description of the addition above, we considered fetch and store as subtasks of it. This of course depends on machine design. On CRAY, the task of routing data from memory and back is assigned to specialized pipelines (in fact, only one such exists on CRAY-1, the "loader" which fetches or stores alternatively; it was rapidly realized that one was not enough to avoid bottlenecks at this level, and this design mistake has been corrected in the XMP series).

This leads us to the concept of "chaining", which is close to the mathematical concept of composition of operations. Suppose one wishes to perform the vector extension of $\theta o \oplus$, where $\oplus$ and θ are distinct scalar operations, taking m_1 and m_2 clocks respectively.

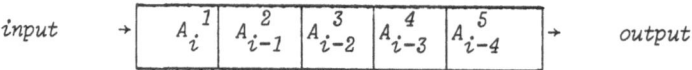

input → $\boxed{A^1_i \mid A^2_{i-1} \mid A^3_{i-2} \mid A^4_{i-3} \mid A^5_{i-4}}$ → *output*

Fig. 4. Tasks which are simultaneously active in the different segments of a pipeline.

84

```
┌─────────────────┬──────────────────────┐   ┌──────────────────────────────────────┐
│ 1   2   3   4   │ n+4  n+5  n+6         │   │ 1   2   3   4   5   6   7             │
│ 2   3   4   5   │ n+5  n+6  ...         │   │ 2   3   4   5   6   7   8             │
│ 3   4   ... ... │ n+6  ...  ...         │   │ 3   4   5   ...                      │
│ 4   ... ...     │ ...                   │   │ 4   ...                              │
│ .....           │ ...           2n+3    │   │ ... ...              ...       n+4   │
│ .....           │ ...    2n+3   2n+4    │   │ ... ...         ...     n+4    n+5   │
│ n  n+1 n+2 n+3  │ 2n+3  2n+4   2n+5     │   │ n   n+1 n+2  ....  ...         n+6   │
└─────────────────┴──────────────────────┘   └──────────────────────────────────────┘
            (a)                                                 (b)
```

Fig. 5. Scheduling for the composition of two vector
operations. Left, in two chimes. Right, in one,
by chaining.

Instead of the scheduling: first $V\Theta$, then $V\Theta$, one can
adopt the one suggested by Fig. 5b. Physically, this
amounts to placing the $V\Theta$ pipeline just at the exit of
the $V\Theta$ one, thus making a new pipeline of length m_1+m_2.
But of course the number of actually installed pipelines
has to be kept small, so this pipeline reconfiguration
is done by the software. Any vector extension can in
theory be performed by a properly designed chain of pipelines. So its execution time would be $m_1+\ldots$ (the
total length of the pipelines) plus $n-1$, i.e. within our
notation, $I = m_1+\ldots+m_k-1$ and $V = 1$. There is always some
mechanism which allows the compiler to chain pipelines
the best it can. If $V = 1$ is achieved, one says the
extension is completed in "one chime". But this is
generally not possible. The loader, for instance, should
as a rule be used twice in most vector extensions, once
to fetch, once to store. This scarcity of available
resources prevents the compiler to chain completely, so
most vector extensions will in fact be executed as a
succession of a few operations, a new pipeline being re-
configurated for each of these. The observed V, i.e.
the number of chimes, is thus most often 3 or 4 (see
Fig. 2).

Let us abandon chaining for the time being and
summarize the three possible implementations of an
extension. First the serial solution:

$$\text{for } i := 1 \text{ to } n \text{ do} \atop \quad \text{for } j := 1 \text{ to } m \text{ do} \atop \qquad A_i^j \tag{16}$$

85

next the way an array of processors would do it:

$$
\begin{array}{l}
\underline{\text{for}}\ j := 1\ \underline{\text{to}}\ m\ \underline{\text{do}} \\
\quad \underline{\text{for all}}\ i\ \underline{\text{in}}\ [1\!:\!n]\underline{\text{do}} \\
\qquad A_i^j
\end{array}
\qquad (17)
$$

and finally the pipeline mode:

$$
\begin{array}{l}
\underline{\text{for}}\ i := 1\ \underline{\text{to}}\ n + m - 1\ \underline{\text{do}} \\
\quad \underline{\text{for all}}\ j\ \underline{\text{in}}\ [\max(\overline{1},i-n+1)\!:\!\min(i,m)]\ \underline{\text{do}} \\
\qquad A_{i-j+1}^j
\end{array}
\qquad (18)
$$

1.2.3. Memory access, regular storage. Multiplication and segmentation are not enough to build a fast machine. Both provide a fast execution mode, but will data transfers match these speeds? In the present state of technology, extracting the content of a memory location and channelling it to the entrance of a functional unit will take much more time than what is needed for one of the tasks A_i^j. Let this time be the clock-period, and suppose extraction from memory takes k CPs, k = 4 being typical.

The way to achieve the desired speed is by "interleaving" banks of memory. At each CP, each bank, in turn, is requested to deliver a data. If there are k such banks, and if the data which must be fetched in succession are in different banks, the speed will be one data-item per CP. So vectors of data should be stored in the orthogonal arrangement suggested by Fig. 6.

Functionally, a fetch can thus be considered as a succession of k elementary operations, plus k' for the travel of the impulsions. Everything goes as if data were delivered through a pipeline of length k + k' (11 on CRAY-1). This pipe is the

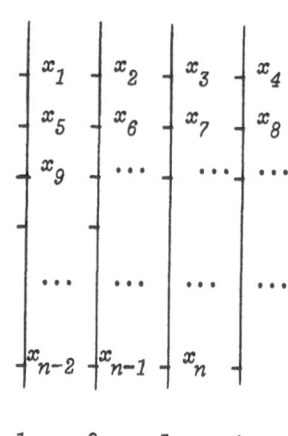

Fig. 6. Storing the components of a vector in a 4-banks interleaved memory.

"loader" we alluded to. As there is only one memory path
on CRAY-1, we shall consider only one such loader is
available, and must serve for fetch and store in alter-
nance.

Now comes the most important point. To fetch a
vector, in order to execute a vector operation, means
picking the components in the right order and putting
them into the loader. So their addresses in core have
to be computed by a hardware element called "address
generator" from parameters describing the vector (Kogge,
1981). If one wants to access regularly spaced elements,
as in the example below:

          DO 1 I = 1, N
      1   A(I) = B(I) + C(3 ✱ I - 2)   (19)

the task of this generator is easy. It goes differently
in the following case:

          DO 1 I = 1, N
      1   A(I) = B(I) + C(ADR(I))   (20)

Unless the address generator equals in complexity
the computing unit itself, it won't be able to do the
job in the case of (20), so (20) is executed in serial
mode. This is a serious problem, as (20) is a rather
common operation in scientific computations, and in the
next few years, adequate hardware, using interconnection
networks, will be available to channel the data at the
right speed even in the case of (20). But for the time
being, we have to take this constraint into account in
our design of the abstract vector machine.

Therefore, only <u>sets of data regularly scattered
through the memory</u> will be considered as vectors. A
data set which is, in the programmer's mind, a vector,
but whose addresses are not in arithmetic progression,
does not satisfy this "storage constraint" we have just
edicted, and thus cannot be considered as a vector in the
sense of the virtual machine we try to describe.

Consequently, <u>no operation can be a vector one unless its output is regularly stored</u> when its input is.
To be specific, consider a permutation π of the integer
segment [1:n] and let p : VECTOR → VECTOR be the induced
operation on vectors:

$$(p(v))_i = v_{\pi(i)}$$ (21)

The programmer could very well decide to store v and p(v) in the same Fortran array, or in two similarly organized such arrays, while keeping the $\pi(i)$ in another integer array. All subsequent operations on p(v) would thus imply indirections, just as in (20), and thus would fail to achieve vector efficiency. Therefore, we cannot accept p as a vector operation of the virtual vector machine, in spite of the conceptual advantage this would give (think about sorting, Gaussian elimination with pivoting, etc.) in many occasions.

We are not through with the address generator. If this is really unsophisticated, it won't be able even to manipulate several vector operands at the same time, for instance when their strides are different (as in (19)). This is the case on CRAY-1. The solution to this problem (and to others, it is a choice in architecture which has other motivations besides this precise one), has been to provide local buffers, called vector registers. The first vector operand is thus fetched and buffered. Then the second one is fetched and this time, in the case of a binary operation, pairs of operands similarly indexed enter the pipeline in succession, coming from two origins: the vector register and (via the loader) the core memory. The output of the pipe is also put provisionally in a vector register. There it may have to wait until the pipelined unit necessary for the next operation is ready (this is the case if this next operation is a store, since the loader is engaged). On the other hand, if this unit is free, the output may serve immediately as input for the next operation. In this case it is still put into the register but simultaneously copied and routed towards the pipeline entrance. This is how chaining is done in practice.

This mechanism is illustrated in the case of addition by the following program (x, y, z are vectors):

$\{x \leftarrow y + z :\}$
    load y in vector register VR1 ;
    load z into VR2 ;   (22)
    VR3 ← VR1 + VR2 ;
    store the contents of VR3 in x ;

and in graphic form by Fig. 7. From the point of view we adopt here, (22) is the implementation of a single vector operation. It could thus theoretically be performed in one chime, but is actually done in 3 (Fig. 2). This is because the unique loader is required three times, and only the second time can it chain to another pipeline

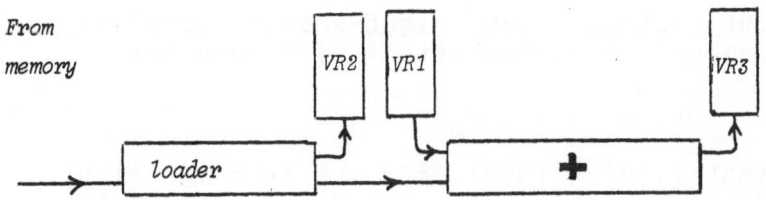

From memory

VR2 VR1   VR3

loader   +

Fig. 7. Active units during the execution of lines 2 and
3 of (22).

(the adder). The Fortran programmer has very little
control on chaining (though by a clever arrangement of
expressions, one can sometimes induce the compiler in
chaining operations that normally it would not chain).
This is why vector operations are considered "from memory
to memory" in our modelling. The CAL programmer (CAL is
the Cray Assembly Language) sees a completely different
virtual machine, as (22) suggests.

Step by step, we have obtained explanations for all
the particularities of Fig. 2, but two: the V of operations which involve reductions (that will have to wait)
and the figures in column R, of which nothing was said
up to now. R is the "reinitialization" time and corresponds to the discontinuity observed at vector length 64
in Fig. 1. It exists because vector registers can
contain exactly 64 scalar elements. For longer vectors,
each new slice of 64 is treated as a new vector, hence
an extra start-up time, only shorter than I, for this
also includes all various overheads.

1.3. The "Vector Machine" Concept

We wish to introduce a new type of program object,
called VECTOR, in addition to the predefined (in Fortran)
types like INTEGER, LOGICAL, etc., all scalar objects.
To avoid tedious distinctions, SCALAR will generically
denote objects of this kind, and we shall not distinguish
between REAL-VECTOR, INTEGER-VECTOR, etc.

A type is defined by its possible value (here
sequences of SCALARs of any finite length), but also,
and much more importantly, by the operations which are
allowed on it, or which relate it to preexisting types.
We defined already length : VECTOR → INTEGER (this symbol-
ism gives the "type" of the operation), abridged in $|u|$,

if u is the vector, τ (the right-shift), broadcast, and the extensions. A left-shift τ' will come handy:

$$\tau'u = \{u_2, u_3, \ldots, u_n\} \tag{23}$$

(remark that $|\tau'u| = |u|-1$, $\tau'\tau = 1$ (the identity), $\tau\tau'$ is a projector which cancels the first component). Let us also introduce component : VECTOR × INTEGER → SCALAR, self-explaining, and its natural abbreviation u_i. (This is an example of a function only partially defined, its domain is $\underline{dom}$(component) = $\{\{u,i\}|\ 1 \leq i \leq |u|\}$.)

Is that all? What should be clear following our previous description, and forgetting the particularities of the CRAY-1, is that <u>only operations which respect the</u> <u>"regular storage" constraint</u> can join the vector extensions in the definition of the vector machine. This is a very strong constraint, which bars practically anything else than the extractions we now define.

The definition we give is a little far-fetched, but this will help avoid heavy notation, and on the other hand, a glance at Fig. 8 should be enough to guess what we are after. First let us choose an integer r > 1, the radix, understood but not explicitly written in what follows. Now, given a vector u, $E_i u$ is the vector of components:

$$(E_i u)_k = u_{i+(k-1)r} \quad \forall\ k,\ 1 \leq k\ \underline{and}\ i+(k-1)r \leq n \tag{24}$$

and $E^j u$ is

$$(E^j u)_k = u_{k+(j-1)r} \quad \forall\ k,\ 1 \leq k\ \underline{and}\ k+(j-1)r \leq n \tag{25}$$

E_1:		1	5	9	13	17
E_2:		2	6	10	14	18
E_3:		3	7	11	15	
E_4:		4	8	12	16	
		E^1	E^2	E^3	E^4	E^5

Fig. 8. Extractions for radix r = 4.

The radix 2 is especially important, and special abbreviations are useful in this case. E_1 is denoted odd, or simply O, because it amounts to picking up the components of odd rank. Similarly, E_2 is even, or E. As one can see, $\forall$ u $\in$ VECTOR,

$$|\text{O } u| = \lceil|u|/2\rceil \qquad |\text{E } u| = \lfloor|u|/2\rfloor \qquad (26)$$

(i.e. the integers just above or just under the half-length).

The idea of the vector machine is to make it suitable for programming in terms of its operations. Therefore, the operation set should have enough algebraic properties. In particular, some invertibility is necessary for all operations. The operation which reconstructs a vector u from its slices $E_i u$ is r-alternate (its arity is r), or simply alt. The similar one for the E^js is r-concatenate, or conc, or // in infixed notation. Let us just explicit alt for r = 2. Its domain is

$$\underline{\text{dom}}(\text{alt}) = \{\{u,v\}|\ |v| \leq |u| \leq |v| + 1\} \qquad (27)$$

and, by definition,

$$\text{O alt}(u,v) = u, \quad \text{E alt}(u,v) = v \quad \forall\{u,v\} \in \underline{\text{dom}}(\text{alt}) \qquad (28)$$

The machine also contains the two vector constants 0 and 1. For technical reasons (one may wish to add or multiply vectors of different lengths), we have adopted the convention $|u+v| = \max(|u|,|v|)$, and $|u{\times}v| = \min(|u|,|v|)$. Consistently, $|0| = 1$ and $|1| = \infty$ (the unique vector of infinite length). Notice that broadcast(0,n), a vector of length n, is not the same as 0.

Figure 9 gives a synoptic of the vector machine. As an algebra, the vector machine has a rather poor structure, nothing like a ring or a field. Nevertheless, it is complete enough to allow program development, as we shall see.

We direct attention to the following algebraic properties, which will prove useful:

$$\text{O}\tau = \tau\text{E}, \quad \text{E}\tau = \text{O}, \quad \text{E}\tau' = \tau'\text{O}, \quad \text{O}\tau' = \text{E}. \qquad (29)$$

NAME AND TYPE OF THE OPERATION		ABBR.	PROPERTIES
zero, one	→ VECTOR	0, 1	vector constants
length	VEC → INT	$\lvert u \rvert$	$\lvert 0 \rvert = 0$, $\lvert 1 \rvert = \infty$
add	VEC × VEC → VEC	$u + v$	$\lvert u + v \rvert = max(\lvert u \rvert, \lvert v \rvert)$
multiply	VEC × VEC → VEC	$u * v$, uv	$\lvert u*v \rvert = min(\lvert u \rvert, \lvert v \rvert)$ $\lvert u*0 \rvert = 0$, $\lvert u*1 \rvert = \lvert u \rvert$
hom	SCAL × VEC → VEC	$\lambda\,u$	$1\,u = u$
abs, sin, −,...	VEC → VEC		extensions of unary operators
sup, inf, ...	VEC × VEC → VEC	∨, ∧,...	extensions of binary operations
right shift	VEC → VEC	τ	$(\tau u)_{i+1} = u_i$, $(\tau u)_1 = 0$
left shift	VEC → VEC	τ'	$(\tau' u)_{i-1} = u_i$
odd	VEC → VEC	O	$E\tau' = \tau'O$, $E\tau = O$
even	VEC → VEC	E	$O\tau' = E$, $O\tau = \tau E$
alt	VEC × VEC → VEC	alt	$O\,alt(u, v) = u$ $E\,alt(u, v) = v$
broadcast	SCAL × INT → VEC		$(broadcast(x, n))_i = x \; \forall\, i \le n$

Fig. 9. The vector machine, operations and properties.

VECTOR ALGORITHMS

We shall now describe a few popular vector algorithms within our notation.

2.1. Linear Recurrence

Let two vectors a and b of length $|a| = |b| = n$ be given. One looks for a vector x, $|x| = n$, such that

$$x - a \times \tau x = b \qquad (30)$$

i.e., using coordinates, $x_1 = b_1$ and

$$x_i = a_i \times x_{i-1} + b_i \quad \underline{for} \ i > 1 . \qquad (31)$$

The scalar type of the components is left unspecified (provided + and $\times$ make sense), so (30) contains in fact a vast class of problems. A special case is a = broadcast(λ,n): then the last component x_n is the value $B(\lambda)$ of the polynomial whose coefficients are the b_is (for (31) reduces to Horner's rule in this case). A still more special case is $\lambda = 1$ (all a_is equal to 1): it corresponds to computing the <u>sum</u> of the components of b.

Of course, with conventional computers, nobody would see any "problem" in (30)! But on a vector one, one feels that the straightforward coding of (31) leaves something to be desired,

$$\underline{for} \ i := 2 \ \underline{to} \ n \ \underline{do}$$
$$\qquad x_i \leftarrow a_i \times x_{i-1} + b_i \qquad (32)$$

since this would be executed at serial speed only. The loop (32) does <u>not</u> translate the vector extension of any operation, so there is no obvious way to make use of a pipeline in this case. If one subdivides the add-and-multiply into more elementary operations as we did before, the dependence graph of (32) is as in Fig. 10.

The point we want to make is this: by coding (32), one departs from the "vector programming discipline"; on

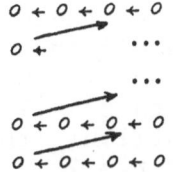

Fig. 10. Dependence graph for (32).

the opposite, if one sticks to it, by making the best effort to avoid invoking other operations than vector ones, vector algorithms will be generated; and this can be done by simple algebraic manipulations on the static statement (30), which is the specification of the problem.

2.1.1. "Iterative" approach: recursive doubling. Let us apply the operator $1 + a\tau$ to (30). This operator consists, when applied to a vector u, in right-shifting u, multiplying by a, then adding u to the result. (Recall that $a\tau u$ has the same length as a, the shorter one). We obtain, since $\tau(a \;\times\; \tau u) = (\tau\ a) \;\times\; \tau^2 u$,

$$x - (a \;\times\; \tau a) \times \tau^2 x = b + a \;\times\; \tau b \tag{33}$$

which is similar to (32), but the shift on u is now τ^2, two positions instead of one. This immediately suggests an iterative method:

> **in** a,b : VECTORS, **out** x : VECTOR$\{|a| = |b| = |x| = n\}$
> **variable** k : INTEGER ;
>
> k ← 1 ;
> **while** k < n **do**
> | b ← b + a $\;\times\; \tau^k$b ;
> | a ← a $\;\times\; \tau^k$a ; $\hspace{3cm}$ (34)
> | k ← 2 $\;\times\;$ k ;
> | $\{x - a \;\times\; \tau^k x = b$, loop invariant$\}$
> x ← b

All the operations in (34) are vector ones. If the programming language was rich enough to incorporate such operations, we would be done. In Fortran, it will be necessary to further refine this program, coding the vector operations as DO-loops on indices. The compiler will check that these loops do code vector extensions of scalar operations, and organize the pipelines accordingly.

We shall refrain here to give a Fortran version of (34) (which is more delicate to develop than it may appear, for there are many possible variants), for any coding would rather obfuscate the structure than help understand it. On (34), we count p executions of the loop, with $p = \lceil \log_2 n \rceil$. To simplify the discussion, let a= broadcast(1,n) (this corresponds to the summation of the b_is) and $n = 2^p$. All further comparisons will be restricted to that case. As there are p vector operations, their total length being $np + 1 - 2^p$, the execution time will be, in CP,

$$I \log_2 n + V (1 + n(\log_2 n - 1)) \qquad (35)$$

instead of $(n - 1)S$ for the standard serial program (32). Algorithm (34) appears in Stone (1973) under the name "recursive doubling".

2.1.2. "Recursive" approach: cyclic reduction. We now turn to a very productive algorithmic principle, the "odd-even" approach, also known as "divide and conquer", etc. Let us simply see what becomes of (30) when we let odd and even act on it:

$$Ox - (Oa) \ast \tau (Ex) = Ob \qquad (36)$$

$$Ex - (Ea) \ast Ox = Eb \qquad (37)$$

One can eliminate either Ox or Ex, which yields

$$Ox - O(a \ast \tau a) \ast \tau (Ox) = O(b + a \ast \tau b) \qquad (38)$$

$$Ex - E(a \ast \tau a) \ast \tau (Ex) = E(b + a \ast \tau b) \qquad (39)$$

Both equations are similar to (30), but concern vector unknowns of about twice shorter length. So we have reduced the problem to one which is "less complex", in an obvious sense, but similar, and this is the paradigm of recursive solutions. Two equivalent possibilities exist, let us proceed with the first one, thence the recursive procedure:

```
program linrec(in a, b : VECTORS, out x : VECTOR
                              {|a| = |b| = n})
    if n = 1 then
        x ← b
    else                                            (40)
        linrec(O(a⋇τa), O(b + a⋇τb), x) ;
        x ← alt(x, Eb + (Ea)⋇x)
```

This program can be transformed into a non-recursive one the standard way, by using a stack of VECTORs:

```
program linrec(in a, b : VECTORS, out x : VECTOR)
    {non-recursive version, using a stack of VECTORs}
    while |a| > 1 do
        put Eb and Ea on top of the stack ;
        a ← O(a⋇τa) ; b ← O(b + a⋇τb)
    x ← b ;                                         (41)
    while |x| < n do
        remove Ea and Eb from the stack ;
        x ← alt(x, Eb + (Ea)⋇x)
```

The operation count, in the same conditions as above, is

$$(2 \log_2 n - 1) I + (2(n - 1) - \log_2 n)V \qquad (42)$$

2.1.3. Discussion. The relative efficiency of the three methods depends on the two dimensionless parameters I/V and S/V. Fig. 11 shows timing results on a CRAY-1, for cyclic reduction and for the serial method. The "iterative" method (34) was not considered, for it can never beat cyclic reduction on the CRAY-1; it requires fewer loop start-ups, but much more arithmetic, and the ratio I/V is too small, on a CRAY at least, to compensate for this. The break-even point between the serial method and the recursive one is approximately at n', the solution of the following equation:

$$2 \log n = n I/S \qquad (43)$$

which shows clearly the importance of the ratio I/S. The larger it is, the longer the vectors should be to gain something with cyclic reduction. Even so, the speed-up is modest. The interest of cyclic reduction is mainly theoretical.

Some people find it helpful to see the dynamics of the execution of a program, in graphic form. Two such

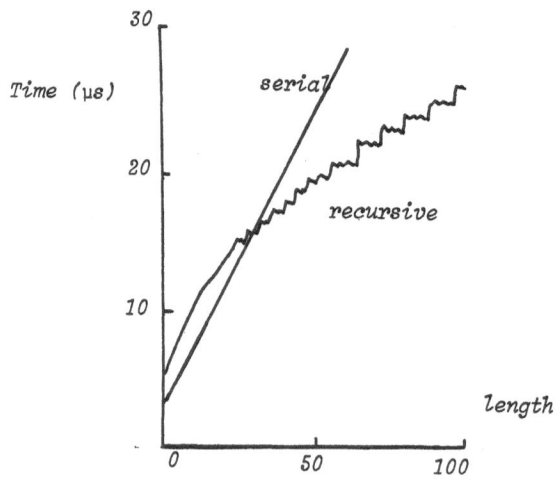

Fig. 11. The recursive method (41) vs serial execution.

96

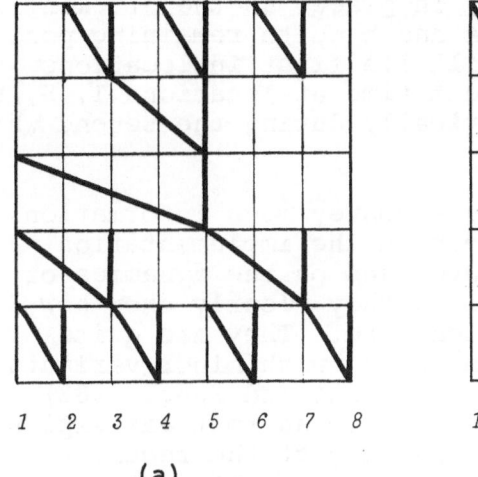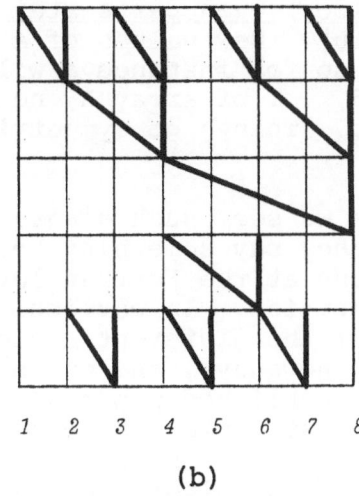

Fig. 12. Sketch of the succession in time of scalar actions for program (41) (left) and the variant in
which (39) is used instead of (38) (right).
Here, n = 8.

diagrams are given in Fig. 12; the one on the left
concerns (41) (elimination of the even parts, recursive-
ly), the one on the right is about the twin algorithm
that one would obtain by eliminating odd parts. The
interpretation of these diagrams necessitates some preliminary explanations.

Before execution, the data a and b lie in Fortran
arrays A, B. There is also an array X which will contain
x at the end. These 3 parallel arrays are indexed by i,
ranging from 1 to n. Columns of the diagram correspond
to i, from left to right. Horizontal lines correspond
to time. More precisely, the interval between two horizontal lines corresponds to one step of either loop in
(41). When two heavy lines, coming from (j, t - 1) and
(k, t - 1) respectively, converge at (i, t), it means
that array locations of index i are affected by some
treatment done on the contents òf locations j and k
between time t - 1 and time t.

Diagram 12a has been obtained from program (41) by
using the following implementation rule (one among many
other possible ones): During the descending phase (first

while loop) even parts stay in place, so the stacking is "in situ". New values of a and b go to remaining positions, so for instance a will lie first in locations
1, 3, 5, ... of array A, next time at locations 1, 5, 9,
... etc. Things go symmetrically during the second while loop.

As we see, such diagrams convey much information, since they say something both on the implementation decisions at the Fortran level and on the dynamics of the execution. In particular, they clearly show how parallel the different actions are. They are quite popular nowadays, and can be seen, with minor variants, in many published papers (e.g. Brent and Kung, 1982). But we must confess some scepticism as to their explana-tory qualities. Does Fig. 12a suggest the recursive nature of the algorithm? Perhaps yes, but program (40) is certainly better in this respect. There is something here which is reminiscent of the old habit of describing algorithms with the help of flow-charts.

2.2. Linear Algebra

Let us display a few algorithms devised for the vector machine. First the often discussed multiplication of a vector by a matrix:

$$
\begin{aligned}
&\underline{\text{program}}\ \text{matrix multiply}\ (\underline{\text{in}}\ a:\ \text{MATRIX},\ x\ :\ \text{VECTOR}, \\
&\qquad \underline{\text{out}}\ y\ :\ \text{VECTOR}\ \{\text{order}(a)\ =\ \text{length}(x)\}) \\
&\qquad y \leftarrow 0\ ; \\
&\qquad \underline{\text{for}}\ j\ :=\ 1\ \underline{\text{to}}\ n\ \underline{\text{do}} \\
&\qquad\qquad y \leftarrow y + x_j \times a^j
\end{aligned}
\tag{44}
$$

(To be completely correct, we should have given a defini-tion of MATRIX, as a type, first. This is rather tedious, and suffice to say that MATRIX objects relate to pre-defined types through operations like

$$
\begin{aligned}
&\text{order}\ :\ \text{MATRIX} \rightarrow \text{INTEGER} \\
&\text{column}\ :\ \text{MATRIX} \times \text{INTEGER} \rightarrow \text{VECTOR} \\
&\text{row}\ :\ \text{MATRIX} \times \text{INTEGER} \rightarrow \text{VECTOR}
\end{aligned}
$$

and that a^j and a_i are abbreviations for column (a,j) and row (a,i).

Note that (44) is the only natural way to write a matrix-vector multiplication on the abstract vector machine, since dot-products are not part of it. The Fortran code is

```
      DO 1 I = 1, N
           Y(I) = 0
C          -- INITIALIZATION OF Y
1     CONTINUE
      DO 2 J = 1, N                                    (45)
           DO 2 I = 1, N
2               Y(I) = A(I,J) ⨉ X(J) + Y(I)
```

and differs from the usual one by the order of the indices
I and J. The inner loop 2 here is the multiplication of
a vector by a scalar, as if J was inside and I outside,
it would be a dot product, with the undesirable overhead
already noticed. This is a trivial but standard example
of how "thinking vectors" leads to better coding in a
natural way.

Similarly, triangular systems can be solved "by
columns":

program triangular system (in a : MATRIX, b : VECTOR,
    out x : VECTOR {a is inf-triangular and regular})
              variable xj : REAL ;
    x ← b ;
    for j := 1 to n { = order (a) = length(b)} do
        xj ← x_j/a_j^j ;   (46)
        x ← x - xj ⨉ a^j ; x_j ← xj

To present Cholesky factorization, we need to introduce
a new operation again,

    × : VECTOR × VECTOR → MATRIX

which is defined by

$$(u \times v)_i^j = u_i \times v_j \tag{47}$$

The development of the algorithm is in two steps. First
notice that what we look for is an inf-triangular matrix
s such that

$$\Sigma_{j := 1 \underline{to} n} \; s^j \times s^j = a \tag{48}$$

and this specification immediately suggests an algorithm:

```
program Cholesky (in a : MATRIX, out s : MATRIX)
    variables pivot : REAL, c : MATRIX ;

    c ← a ;
    for j := 1 to order(a) do
        pivot ← sqrt(s_j^j) ;                                    (49)
        s^j ← c^j/pivot ;
        c ← c - s^j × s^j
```

Now let us get rid of the slack variable c and develop
the dyadic product ×, in order to be at the level of the
vector machine again:

```
program Cholesky (in a : MATRIX, out s : MATRIX)

    s ← inf-triangular part of a ;
    for j := 1 to order(a) do
        pivot ← sqrt(s_j^j) ;                                    (50)
        s^j ← s^j/pivot ;
        for k > j do
            s^k ← s^k - s_k^j ×· (Π_k s^j)
```

(We have omitted the definite-positivity test for simplicity; Π_k is the projector on the last n - k components.)

Again this factorization "by columns" is less
familiar than the one "by rows", and the point to make
is that (50) comes naturally from the specification (48).
State the problem in vector terms, the vector method
will follow. And the role of the "vector machine"
concept is to help stating the problem "in vector terms".

2.3. The "Total Reduction" Problem

We now come back to the problem of solving recurrences, but in a different context. The scalar type will
have a weaker structure, but on the other hand, we shall
get more insight on algorithms thus reduced to their
most basic features.

2.3.1. Statement of the problem.
Definition: A
monoid is a set equipped with an associative binary
operator o with identity (i.e. there exists an element
e such that $e_o x = x_o e$ for all x).

Given a monoid, we shall call SCALARs its elements.
We define, as above, VECTORs as finite sequences of

100

SCALARs. The definitions of length and of the extraction operators, as given in § 1.3, (24) and (25), are left unchanged. We also retain the definition of the shifts τ and τ' ((8) and (23), but what was previously called "0" is now the identiy e, so

$$\tau u = \{e, u_1, \ldots, u_n\} . \tag{51}$$

Since there is only one scalar operation, namely $\circ$, there is a unique vector extension, still denoted by $\circ$, and defined by

$$u_\circ v = \{(u_\circ v)_i | i = 1, \ldots, \min(|u|, |v|)\}$$

Remark that $|u_\circ v| = \min(|u|, |v|)$. This confers a monoid structure on VECTORS, the identity being the vector broadcast (e,1), denoted by e.

On this utterly simplified machine, there is only one problem of interest: Given a vector a of length n, compute optimally x such that

$$x = a_\circ \tau x , \tag{52}$$

i.e., in coordinates,

$$x_i = a_i \circ a_{i-1} \circ \cdots \circ a_1 \equiv a_i \circ x_{i-1} \underline{\text{if}} \ i > 1. \tag{53}$$

This is called the <u>total reduction</u> problem, when all components of x are needed, not only the last one (that would be the reduction problem, according to the definition given in (15) above). Our notation is

$$x = \text{red}(a) \tag{54}$$

This problem is really fundamental, for the monoid structure, being rather weak, can be recognized behind many situations. The first, obvious, example, is finding the sum (or the product) of the components of a real vector: this is the reduction problem, with $\circ = +$ (or $\times$). There is a great variety of other examples, for which we now give a general, all-encompassing presentation.

<u>2.3.2. Examples of problems of the "total reduction" family</u>. Let T and S be two sets, and

$$f : T \times S \rightarrow S \tag{55}$$

a given function. Consider the monoid (S $\rightarrow$ S) of all

mappings from S into itself, with composition as operator
$\circ$ and the identity mapping as e. To f corresponds a mapping

$$F : T \rightarrow (S \rightarrow S) \tag{56}$$

from T into this monoid, defined by (F_t is the image of
t by F):

$$F_t(s) = f(t,s) . \tag{57}$$

Suppose that, given $s_1 \in S$ and a vector t of n
elements of T, we want to compute the successive values
of

$$s_i = f(t_i, s_{i-1}), \quad i = 2, \ldots, n. \tag{58}$$

This defines a mapping $x_i \in (S \rightarrow S)$: x_i is the application which associates s_i with the starting value s_1.
Let us also call a_i the image of t_i by F.

Now, of course, the problem to be solved is (52).

By solving (52), we shall do more than was asked
for, since mappings from S to S will be found, thus the
recurrence (53) will be solved whatever the initial
value.

The reader, who can guess now about many ways to
solve (52), may wonder: so all non-linear recurrences
can be solved? Of course, there is a catch. All of
this is purely theoretical unless elements of the monoid
which intervene in the vector algorithm can be explicitly represented on the machine.

This is true of all images of t: F_t is represented
by t, and we may assume some implementation exists for
elements of type T. But $F_t \circ F_t$ is of type $S \rightarrow S$ and
may have no convenient representation, even if S-type
objects have one. (For instance, think of S as the
INTEGERs: objects of type $S \rightarrow S$ are integer valued
functions of integers, and there is no really convenient
way to deal with these as program objects; Fortran does
provide FUNCTIONs, but FUNCTIONs cannot be composed, nor
even added or multiplied together to give other FUNCTIONs).

So, in practice, we shall restrict to functions f
such that $F_t \circ F_{t'}$ is always the image by F of some t''.
The mapping $\{t,t'\} \rightarrow \phi(t,t')$ is Kogge's "companion

function" (Kogge, 1981), when it exists: it is thus characterized by

$$f(t, f(t',s)) = f(\phi(t,t'),s) . \qquad (59)$$

All "scalar" elementary operations will thus amount to
evaluating the companion function for given arguments t
and t'.

A celebrated example of non-linear occurrence which
bends to this treatment (and, we fear, the only really
workable one) is

$$x_i = \frac{a_i \times x_{i-1} + b_i}{c_i \times x_{i-1} + d_i} \qquad (60)$$

because of the group structure of homographic transforms.
Here, S is the REAL type, T is the set of 2×2-matrices.
By rewriting (60) as $x_i = y_i/z_i$ and

$$\begin{vmatrix} y_i \\ z_i \end{vmatrix} = \begin{vmatrix} a_i & b_i \\ c_i & d_i \end{vmatrix} \begin{vmatrix} y_{i-1} \\ z_{i-1} \end{vmatrix} = t_i \begin{vmatrix} y_{i-1} \\ z_{i-1} \end{vmatrix} \qquad (61)$$

we see that the companion function is

$$\phi(t,t') = tt' . \qquad (62)$$

This is an example where there exists an imbedding of the
image $F(T)$ of T in all mappings from S to S in T itself,
so one can exploit the group structure of T.

A particular case, of course, is the linear recurrence ($c_i = 0$). Also, in this last case, one can generalize: x_i and b_i can belong to a vector space, a_i being a
linear operator. So linear recurrences of any order can
be considered. This is well, but one would like to see
an example of practically interesting problems like (60).

There is one: Consider a symmetric triangular matrix,
with diagonal $\{a_1, a_2, \ldots, a_n\}$ and subdiagonal
$\{b_1, b_2, \ldots, b_{n-1}\}$. We want the Cholesky factorization,
i.e. a bidiagonal matrix with diagonal $\{s_1, \ldots, s_n\}$ and
subdiagonal $\{t_1, \ldots, t_{n-1}\}$. The recurrence to be solved
is

$$t_{i-1}^2 + s_i^2 = a_i, \qquad s_i t_i = b_i, \qquad (63)$$

i.e., eliminating t_i,

$$s_i^2 = a_i - \frac{b_{i-1}}{(s_{i-1})^2} \, , \tag{64}$$

a problem of the form (60), with $x_i = s_i^2$.

2.3.3. Vector methods for problem (52). Let us again start from the specifying equation

$$x = a_0 \, \tau x \tag{52}$$

and let vector operators act on it.

The first available one is τ. Applying it to (52) yields:

$$x = (a_0 \, \tau a)_0 \tau^2 x \tag{65}$$

and the iterative method we have already studied emerges.

What happens with odd and even is no news either:

$$Ex = Ea_0 Ox = Ea_0 (Oa_0 \, \tau Ex)$$
$$= (Ea_0 \, Oa)_0 \, \tau (Ex) \tag{66}$$

thence cyclic reduction, since (66) is to Ex what (52) is to x.

Let now the E_is act on (52) (their definition is in (24)). We get

$$
\begin{vmatrix}
E_1 x = E_1 a_0 \tau (E_r x) & \text{(for } E_1 \tau = \tau E_r) \\
E_2 x = E_2 a_0 \, E_1 x & \text{(for } E_i \tau = E_{i-1} \text{ if } i > 1) \\
\dots & \dots \\
E_r x = E_r a_0 \, E_{r-1} x & \dots
\end{vmatrix}
\tag{67}
$$

whatever the radix $r \geq 2$, and thus

$$E_r x = (E_r a_0 \, \dots \, _0 E_1 a)_0 \tau E_r x \, , \tag{68}$$

again the basis for a recursive solution ("r-cyclic reduction"). So, once $E_r x$ has been found by a recursive application of the same procedure, one finds the other parts of x by

$$E_i x = (E_i a_0 \, \dots \, _0 E_1 a)_0 \tau E_r x \, , \quad 1 \leq i < r \, . \tag{69}$$

So $E_i a_0 \, \dots \, _0 E_1 a$ is needed for all i from 1 to r. While r remains small, this is a matter of r - 1 successive vector operations; (67) also needs r vector operations, and x is reconstituted by the alternation of the E_is. The algorithm so obtained is not fundamentally different from odd-even reduction (2-cyclic reduction).

But if r is large, something new appears. Finding $E_i a_0 \ldots {}_0 E_1 a$ is again a total reduction problem. But it cannot be solved by a recursive call to the same r-cyclic procedure, as one is tempted to propose, for the "components" $E_i a$ are themselves vectors. This is precisely the point where the other family of extraction operators, the E^js (see (25)) become helpful. We just remark that

$$(\text{component } j \text{ of } E_i u) = (\text{component } i \text{ of } E^j u) \; \forall \, u \, \epsilon \; \text{VECTOR} \quad (70)$$

(clear from the definitions (24) and (25)), and therefore,

$$(\text{component } j \text{ of } E_i a_0 \ldots {}_0 E_1 a) = (\text{component } i \text{ of } \text{red}(E^j a)) \quad (71)$$

So each step of r-cyclic reduction consists in solving $\lceil n/r \rceil$ total reduction problems (independent and similar, and in small number if r is large enough), besides the r - 1 vector multiplication (69). But these also transform to:

$$E^j x = \text{red}(E^j a)_0 \; (\tau E_r x)_j, \qquad 2 \leq j \leq \lceil n/r \rceil , \quad (72)$$

by the same trick (70). The last factor on the right is the last component of $E^{j-1} x$, if j > 1, and e otherwise. Finally, introducing the ad-hoc operator,

$$E^j x = \text{red}(E^j a)_0 \; \text{last\_component}(E^{j-1} x) \; \underline{\text{if}} \; j > 1$$
$$= \text{red}(E^1 a) \quad \underline{\text{if}} \; j = 1 . \quad (73)$$

Especially interesting is the extreme case $r = \lceil n/2 \rceil$, for then $\lceil n/r \rceil = 2$, and the corresponding method is dual to 2-cyclic reduction. Just as E_1 and E_2, with radix 2, were denoted odd and even, we shall here, to emphasize symmetry, baptize head and tail, respectively, the operators E^1 and E^2 with radix $\lceil n/2 \rceil$. The homolog of alt (see (28)) is conc, which reconstructs a vector from its head and tail:

$$\text{head}(\text{conc}(u,v)) = u, \quad \text{tail}(\text{conc}(u,v)) = v,$$
$$\forall \{u,v\} \, \epsilon \; \underline{\text{dom}}(\text{conc}) \equiv \{\{u,v\} \mid |u-1| \leq |v| \leq |u|\} (74)$$

Let us, for the sake of comparison, first rewrite the program for 2-cyclic reduction:

<u>program</u> CRred (<u>in</u> a : VECTOR, <u>out</u> x : VECTOR)
> <u>if</u> |a| = 1 <u>then</u>
>> x ← a
>
> <u>else</u>　　　　　　　　　　　　　　　　　　(75)
>> CRred(Ea $_\circ$ Oa, x)
>> x ← alt(Oa $_\circ$ τx, x)

next the dual one:

<u>program</u> RDred (<u>in</u> a : VECTOR, <u>out</u> x : VECTOR)
> <u>variables</u> y, z : VECTORS ;
> <u>if</u> |a| = 1 <u>then</u>
>> x ← a
>
> <u>else</u>　　　　　　　　　　　　　　　　　　(76)
>> RDred(head(a), y) ;
>> RDred(tail(a), z) ;
>> x ← conc(y, z$_0$ last_component(y))

Fig. 13 shows a dynamic diagram for n = 8 (compare with Fig. 12b).

Cyclic reduction with radix 2 and its dual variant are <u>not</u> equivalent, even for powers of two. In fig. 13, we have put on the same horizontal level different vector operations which can be done in parallel. One could

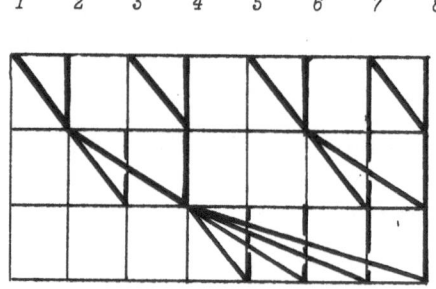

Fig. 13. Dynamic diagram for the "dual method", assuming total reductions which can be done in parallel are. Cf. Fig. 12b. RD and CR are <u>different</u>, but quite close to this case.

think that by similarly exploiting the remaining parallelism in Fig. 12b, this would be transformed into Fig. 13, but this is not so (consider how x_7 is obtained in the two cases). Besides, the identity of the first two horizontal stripes in the two diagrams is only an accident.

Remark. If people were free to rename things their way, I confess I would like to call algorithm (76) "recursive doubling". Kogge and Stone (1973) describe recursive doubling as "a concept [...] which consists of breaking the calculation of one term into two equally complex subterms", and this seems quite fitting when applied to (76). Moreover (76) is recursive (i.e. at least one of the statements of the procedure body is an invocation of this same procedure, only applied to different data), and deserves a name, since it is similar to and irreducibly different from cyclic reduction. But what is called "recursive doubling", following Stone (1973), or Dubois and Rodrigue (1977) is the iterative algorithm (34), and algorithms built from the same patterns, especially for solving tridiagonal systems of equations. It is one of life's ironies that recursive doubling is not recursive on any account.

Remark. Considering that one r-reduction cycle consists essentially in another total reduction, there is no compelling reason to keep the same radix r for the next cycle. Determining the optimal sequence of radices for a given value of n is possible, by dynamic programming, but this seems a problem of no more than academic interest. It generalizes the classical problem of computing a monomial x^n with a minimal number of multiplications (Knuth, 1969).

The last remark suggests something which, on the contrary, seems of appreciable practical value: instead of going all the way through the reduction process, whatever the value of n, one should stop the recursion and revert to serial mode after a few r-reduction steps, as soon as the vectors one works on are too short for further reductions to be competitive. This is common sense. But if so, why not, for the sake of simplicity of programming only one r-reduction, followed by the rest of the computation in serial mode. The problem is to find the best value of r in this context, and one usually chooses $r = n^{1/2}$. This is referred to as "partitioning". The trick forms the basis of the method by Wang (1981) to solve tridiagonal equations. The same idea has been developed by Calahan (1980), who refers to it as "vectorizing across sub-systems".

<u>2.3.4. A pipeline for the reduction problem</u>. What comes now is a little out of the scope of the (abstract) vector machine, since the trick involved depends on the use of segmented functional units. It would not make sense on a multi-processor. On the other hand, the solution we shall describe is in the spirit of what is called "systolic architectures".

Previously, we transformed the basic equation

$$x = a_0 \ \tau x \tag{77}$$

by letting τ act on it. Iterating the process, we get, since τ distributes with respect to $_0$,

$$
\begin{aligned}
x &= a_0 \ \tau a_0 \ \tau^2 x \\
  &= a_0 \ \tau a_0 \ \dots \ _0\tau^{m-1}a_0 \ \tau^m x \\
  &= \alpha_0 \tau^m x
\end{aligned}
\tag{78}
$$

Now, we remark that <u>a pipeline with m segments</u> which performs the binary operation $_0$ <u>can be turned into a machine to solve equation</u>

$$x = \alpha_0 \ \tau^m x \tag{79}$$

<u>in one chime</u>. This is apparent on Fig. 14: the output of the pipeline is fed back into the entrance of the same, so components of the other vector input pair with those of x, only delayed m CPs. (During the first m CP's, the absent input is replaced by a flow of e's, the identity element.) So, all goes as if α and $\tau^m x$ were the operands. The output being x, we thus have a functional unit which specializes in solving equation (79).

So (77) is solved provided α has been computed first. This is done in m - 1 vector extensions of $_0$. These can be chained. One can even, since computing α is nothing

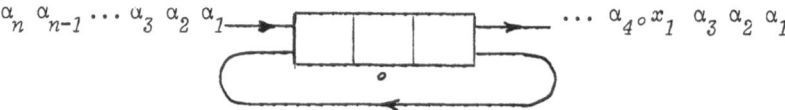

Fig. 14. Self-chained pipeline, in order to solve (79) (3 segments).

else than reduction, apply the iterative method. To be
specific, suppose m = 4. Then (and similarly for all
powers of two):

$$\alpha = (a_0 \ \tau a)_0 \ \tau^2 (a_0 \ \tau a) \tag{80}$$

The operands being each time a vector and the same vector
delayed, the computation can be done by two chained
pipelines, with the help of another piece of hardware:
a "delay latch," whose sole function is to keep a data
item one CP before releasing it unchanged. Similar
generalizations of the pipeline concept are the subject-
matter of Kogge (1981). Figure 15 shows the cascade
solving (77) for m = 4.

2.4. Sum of the Components, and Dot-Product, on CRAY-1

No "cascade" like the one just described exists on
present vector computers (at the vector level; pipelined
adders of similar design were in use long before vector
computers; after all, an integer is nothing else than a
vector of bits, and the carry problem is another instance
of (77)). Nevertheless, the feed-back idea is used on
the CRAY to accelerate the summation of the components
of a vector. In this case, the $_0$ operator is +, and the
key-point is the commutativity of +.

Proposition : If ∘ is commutative, and if y is solution to

$$y = a_0 \ \tau^m y \tag{81}$$

then

$$x = y_0 \ \tau y \ _0 \cdots _0 \ \tau^{m-1} y \tag{82}$$

is solution to (77), the total reduction problem on vector
a.

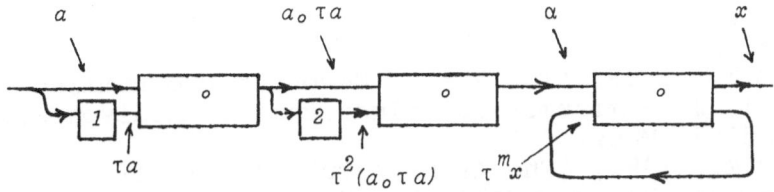

Fig. 15. A cascade of pipelines for (77), with m = 4.

Proof: Let x be defined by (82). Then, by commutativity,

$$a_0 \tau x = a_0 \tau (\tau^{m-1} y_0 \ldots {}_0 y)$$
$$= a_0 \tau^m y_0 (\tau^{m-1} y_0 \ldots {}_0 \tau y)$$
$$= y_0 (\tau^{m-1} y_0 \ldots {}_0 \tau y)$$
$$= x \qquad\qquad \square$$

Since (81) is solved by feed-back, one gets x, in the case m = 4, by a cascade similar to the one in Fig. 15, but with the self-chained pipeline in front of the two others. Only because of commutativity is this permutation allowed. Now, suppose the operator is + and only the last component of x is wanted, which is the case if one wants the sum of the components of a. Then, one needs only the self-chained adder; its output is routed to a vector register, from which the last m components will be extracted and summed in serial mode. This last operation is slow, and we now have at last the explanation of the large "start-up" time (in fact, this time is used at the end of the operation) of the sum and dot-product operations.

REFERENCES

Brent, R.P., and Kung, M.T., 1982, A regular layout for parallel adders, IEEE Trans. on Comp., C31, 3, 260: 64.

Calahan, D.A., 1980, Multi level vectorized sparse solution of LSI circuits, in: "Proc. IEEE Conference on Circuits and Computers", IEEE, New York.

Dijkstra, E.W., 1976, "A Discipline of Programming", Prentice-Hall, Englewood Cliffs, N.J.

Dubois, P.F., and Rodrigue, G.H., 1977, An analysis of the recursive doubling algorithm, in "High Speed Computer and Algorithm Organization", D.J. Kuck et al., eds., Academic Press, New York.

Gries, D., 1981, "The Science of Programming", Springer, Berlin.

Hockney, C.W., and Jesshope, C.R., 1981, "Parallel Computers", Adam Hilger, Bristol.

Knuth, D.E., 1969, "The Art of Computer Programming (Vol. 2/ Seminumerical algorithms)", Addison-Wesley, Reading, Mass.

Kogge, P.M., 1981, "The Architecture of Pipelined Computers", Mc Graw-Hill, New York.

Kogge, P.M., and Stone, H.S., 1973, A parallel algorithm
    for the efficient solution of a general class of
    recurrence equations, IEEE Trans. on Comp., C-22, 8,
    786:93.
Meyer, B., and Baudoin, C., 1978, "Méthodes de program-
    mation", Eyrolles, Paris.
Ramamoorthy, C.W., and Li, H.F., 1977, Pipeline Archi-
    tecture, ACM Comp.Surveys, 61:102.
Robinson, D., 1980, "A Course in the Theory of Groups",
    Springer, New York.
Stone, M.S., 1973, An efficient parallel algorithm for
    the solution of a tridiagonal system of equations,
    J.A.C.M., 20, 1, 27:38.
Wang, H.H., 1981, A parallel method for tridiagonal
    equations, ACM T.O.M.S., 7, 170:83.

III. MIND SUPERCOMPUTERS

MIMD SUPERCOMPUTERS FOR NUMERICAL APPLICATIONS

Ph. Berger, D. Comte, and Ch. Fraboul

Département d'Informatique
ONERA-CERT
BP 4025   31055 Toulouse Cedex, France

INTRODUCTION

The numerical simulation becomes increasingly important day by day in domains which are more and more diverse: aerodynamics, hydrodynamics, structural analysis, nuclear weapons, weather prediction, etc. [1]. The needs in computation power are unmeasurable and far from the possibilities available today [2][3]. The super-computers which are on the market at the present time are the result of technological progress. Although this is important in itself, we have not yet been able to bridge the gap between the needs and the possibilities offered. The supercomputer designers have realized that the solution to this problem lies with the coming of the multiprocessors.

Clustering two, four or more processors on the same program opens up a whole range of new possibilities, but there still remains the problem of mastering these complex architectures. The MIMD multiprocessors totally review not only the sequential programming languages but also whole conventional algorithmic methods designed in a monoprocessing context.

This article discusses a few thoughts both on new algorithmic methods well-adapted to MIMD architectures and the new tools to be put at the disposal of numeric-ians to program these machines.

Having started with vector computers, Chapter 1 rapidly summarizes various supercomputer architectures. In Chapter 2, we shall discuss a few numerical methods in an MIMD multiprocessing environment. Chapter 3 presents a few MIMD architectures studied and evaluated at ONERA-CERT in the context of the French Supercomputer Project called MARIANNE. Finally, Chapter 4 presents a concrete example of a MIND multiprocessor model built at ONERA and gives a few results on its actual performances.

1. VECTOR COMPUTERS - PARALLEL COMPUTERS

1.1. Architectural type

The basic characteristic of a supercomputer is to exploit the parallelism of a program. However, the type of parallelism used can differ from one computer to another, thereby opening up possibilities of various types of architectures.

Vector computers exploit a low level of parallelism between elementary operations. In fact, a floating point operation can be split up into elementary steps (the adding up of the characteristics, calculation of the mantissa, normalization of the result), each are being executed by a specialized hardware operator. As long as these calculations remain sequential by nature, they can operate on different elements of a vector during the processing. This is the principle of the pipeline concept after which pipeline computers have been named. However, the execution of the program advances sequentially, instruction by instruction, exactly the same as on a classical Von Neuman computer.

The vector computers which are currently in use are made up of several pipelined operators (multipipes) which can operate in parallel. For example, if you want to compute the vector expression A$^\times$B+C, you can chain the multiplication and addition calculations on a multipipeline processor; you thereby avoid a superfluous intermediate writing of A$^\times$B result in the main memory.

The array processors (SIMD) exploit a different kind of parallelism. Their processing unit is made up of N identical elementary processors which all execute the same calculation on different vector elements at the same time. Figure 1 shows the listing of existing machines in this category. The execution of the program, as with vector machines, remains sequential.

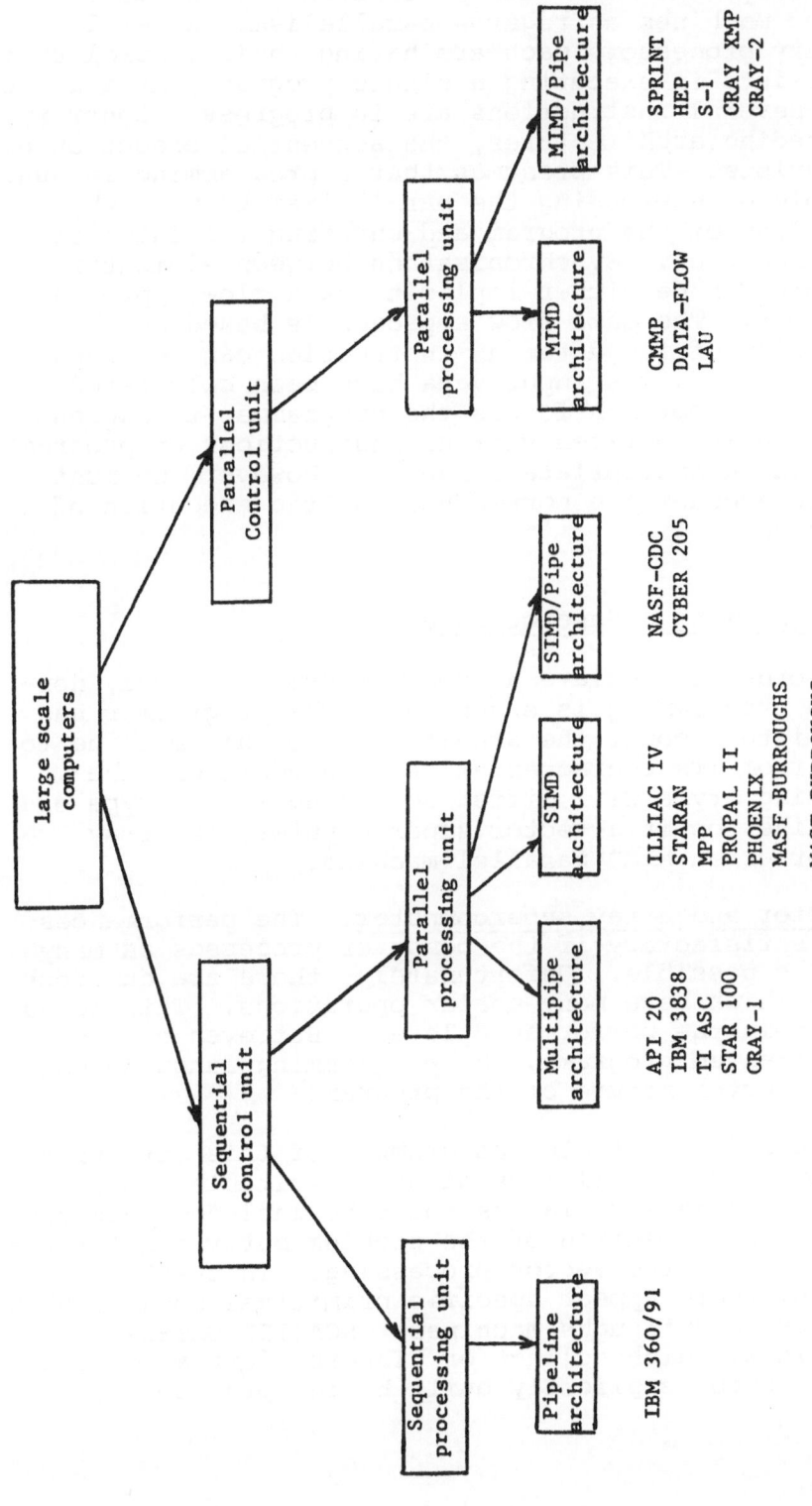

Figure 1. Large scale computer organization

Finally, the MIMD multiprocessors are the most developed machines as regards parallelism. Several elementary processors each are having their control unit, which assists in executing a single program. At a given moment, several instructions are in progress. Contrary to preceeding architectures, the sequential execution no longer exists. This presumes that a programming language is capable of expressing the parallelism between the instructions of the program and ensuring the integrity of the execution. Synchronisation between elementary processors can be either implicit (data flow approach) or explicit. The data flow approach is based on the availability of the data: an instruction can be executed if, and only if, its input data have been calculated. The explicit approach forces the programmer to express the parallelism between various instructions or program tasks with an appropriate language. However, he must take into account the correctness of the execution of the program.

1.2. Programming a Supercomputer

In order to achieve a good performance level, non-standard programming is essential. The programmer has to take into account the architecture of his machine to benefit from its characteristics. In addition, the programming style will differ according to the type of the machine, be it a vector supercomputer, an array processor or an MIMD parallel machine.

Vector and array supercomputer. The performances will be satisfactory if the computer processes as many vectors as possible. Unfortunately, there are numerous programs which have many scalar operations. This leads to a drop in the performance level. But even on the vector-oriented programs, the programming language can hide the vector nature of the program.

Figure 2 illustrates an example of this situation. Vector processing will necessitate a suitable programming: intermediate variables could be included, unnecessary for the resolution of the problem but vital for the acceleration of the vector processing. In current languages, there appear specific primitives such as PACK and UNPACK on Burroughs machines, SCATTER/GATHER on CDC machines, which collect or disperse data according to a bit vector explicitly built by the programmer.

```
WE WANT TO EXECUTE:
   DO 1 I=1,N
   IF (D(I).EQ.0) GOTO 1
   Q(I)= A(I)/D(I)
1  CONTINUE

ON BSP, WE CAN TRANSLATE THAT BY:

   LOGICAL LTEMP(N)
   REAL DTEMP(N), ATEMP(N), QTEMP(N)
   LTEMP(1:N)= D(1:N).NE.0
   PACK/POP=M/WHERE (LTEMP(1:N))
                        DTEMP(1:M)= D(1:N)
   PACK/POP=M/WHERE (LTEMP(1:N))
                        ATEMP(1:M)= D(1:N)
   QTEMP(1:M)= ATEMP(1:M)/DTEMP(1:M)
   UNPACK WHERE (LTEMP(1:N))
              Q(1:N)= QTEMP(1:N)
         OTHERWISE
              Q(1:N)=Q(1:N)
```

Fig. 2. Vector programming example.

```
Initial loop:
   A(0)=1
   LOOP 10 I=1,N
10 A(I)= A(I-1)*B(I)+C(I)

Parallelized loop:
   A(0)= A0
   A(1)= A1
   A(2)= A2
   A(3)= A3
   REP 100 FOR ALL I/(0..3)
      LOOP 10 J=1 STEP 4 TO N-1/4
         K=J*4
         A(K+I)= A(K+I-4)*B(K+I)*B(K+I-1)*B(K+I-2)*B(K+I-3)
                 +C(K+I-3)*B(K+I)*B(K+I-1)*B(K+I-2)
                 +C(K+I-2)*B(K+I)*B(K+I-1)
                 +C(K+I-1)*B(K+I)
                 +C(K+I)
   10 CONTINUE
   100 CONTINUE
```

Fig. 3. Parallelization of a loop on a SIMD architecture.

The programming style can become cumbersome as illustrated in Figure 3. In order to help the programmer, most of the existing compilers accomplish a limited or an extensive amount of the automatic parallelization of the source code [4]. For this, the section concerned must obey many rules. Thus, the CRAY Fortran Compiler parallelizes the iterative loops provided that:

- there is no jump outside the body of the loop,
- there are no subroutine calls,
- there is no defined index reference outside the loop,
- the loop is the most nested one,
- etc. ...

In practice, if the programmer is not careful, only a small proportion of the program satisfies all these rules and the resulting performances are average or disappointing.

Parallel MIMD supercomputers. A parallel MIMD supercomputer is capable of exploiting all types of parallelism:

- Vector parallelism at the elementary processor EP level (if EP are pipelined),
- Scalar parallelism between the different EPs,
- Parallelism between instructions.

As far as programming is concerned, the problem is in expressing these different types of parallelism. There are several possible ways of exploiting such type of architecture:

- implicit parallelism: the programmer does not have to worry about the particular architecture of the machine. He expresses the parallelism of his program in a suitable language (other than sequential Fortran) endeavouring to find the most concurrent algorithm. This parallelism is exploited by the hardware. This approach has not yet been realized on a commercial level and the only prototypes in existence have been made by the research laboratories. Data flow, single assignment and reduction machines are all in this category [5] [6]. The principle of sequentiality is irrelevant here. The program's control execution is fundamentally based on the data and no longer on the location of the instructions in the memory. The languages derived differ radically from a semantic point of view from the classical ones (e.g. Fortran), even if they are fairly similar syntactically. However, with present day technology, it cannot be taken for granted that these machines will be capable of exceeding the numerical supercomputer performances. The control mechanisms are attractive but complicated to execute from the point of view of hardware. On the other hand, their preferred field of application seems to be symbolic computation (artificial intelligence). In this domain, the type of potential parallelism is not as regular as a vector computer and is consequently unsuited to traditional sequential architectures.

- explicit parallelism: this approach involves the splitting up of a program into independent sections called tasks. Each task has to be totally executed on an EP. If the latter have a conventional architecture (sequential, vector or array) the tasks can be written in a sequential language (e.g. Fortran). As well as a purely algorithmic type of processing, the user has to describe the orders of the different tasks to be executed in a specific language. This solution has the advantage of ensuring the continual development of the programming methods (pure Fortran on a monoprocessor to Fortran plus the synchronization language on a multiprocessor). However, as far as the hardware is concerned, the EPs remain conventional. This latter approach has been adopted for the French MIMD Supercomputer Project called MARIANNE.

2. NUMERICAL APPLICATIONS ON MIMD PARALLEL SUPERCOMPUTERS

2.1. Parallel Approach to solve Numerical Problems

In the field of numerical simulation, vector parallelism is now commonly exhibited on supercomputers like CRAY-1, CYBER 205, ... But another complementary kind of parallelism exists in the resolution methods. In order to illustrate this unusual parallelism, we shall give some examples which have been studied in our company.

a) In the domain of aerodynamics, the 3D flow with vortex sheets study can lead to the examination of the evolution of vortices issued from a wing tip [7]. The interdependence of these vortices allows the exploitation of an MIMD parallelism and not vector parallelism (study of the simultaneous progression of each of the eddies).

b) Continuing in the domain of aerodynamics, the resolution of a transonic flow problem using finite difference approximations can reveal, during a convergence iteration, four sub-domains on an initial grid (i.e. subsonic regions, sonic, shock and supersonic). If the resolution scheme is explicit, four different formulas can be applied to each point according to its nature. The dimension and position of these sub-domains are also variable during the iteration [8]. Parallelism between the sub-domains is MIMD in two ways: being a different algorithmic, and having variations in the sub-domains definition.

c) In structural analysis, the size of the data used often forces the user to develop techniques of sub-structuration. The general structure is split up into sub-structures which are then coupled according to a binary tree (eliminating the internal unknowns) so as to reduce the size of the initial problem. As the sub-structures are different (in size, geometry and composition) only a MIMD parallelism allows a simultaneous descent in the various branches of the binary assemblage tree.

d) More generally speaking, a resolution of a part-ial difference equation by finite element method, implies an assembly phase in which local contributions to a given element are added onto a data structure re-presenting the state of the system (matrix or vector). The calculation of the elementary contributions can often be vectorized but the updating phase is inherently sequential: two contributions coming from adjacent finite elements can modify the same coefficient belonging to the global data structure. MIMD parallelism can be found between the calculation of the contributions and the up-dating: it is a multi-producer/one-consumer asynchronous mechanism with an interface made up of buffers allowing the collection of the contributions [9].

e) A more specific type of sub-structure philosophy rests on the principle of nested dissections [10][11]. When applied to discretisation using the finite element method, this method splits up the unregularly formed grid into dissimilar sub-domains (there is also a difficulty in defining optimum separators).
As in the case of the sub-structures mentioned previous-ly, MIMD parallelism can be found in the partial reso-lutions of linear subsystems, whose numbering of unknowns is the result of nested dissections.

f) The resolution of a linear system by block relaxation method can prove to be very efficient on an MIMD system if the numerical context allows for the use

of chaotic or asynchronous iterations [12] [13]. It is
thus the elementary processing speed (hardware level)
which determines the number of required iterations on
the different blocks in order to advance towards
convergence (numerical level).

g) The development of a direct resolution method by
factorization for a given problem is generally followed
by a back- and forward-substitution phase to solve two
triangular problems one after the other. A splitting up
into square or rectangular blocks of the matrix representing the linear system induces a splitting of the algorithm
into computing tasks. The data dependences between these
tasks determine a sequencing graph which gives the order
of the execution of the tasks [14] [15]. The optimal
interpretation of this graph implies the use of MIMD
parallelism.

2.2. Programming Aspects on MIMD Supercomputers

In order to develop and exploit these methods, the
user needs to know certain characteristics of his system
(hardware and software). As an MIMD system is multiprocessor, two questions arise: one regarding the multiprocessing aspect, the other the MIMD nature of the
system.

The former is relative to the hardware and the
operating system:

- what is the nature of an elementary processor?
    In order to achieve a good performance level,
    it must be able to rapidly execute a given
    algorithmic task code exploiting internal
    parallelism (e.g. pipelined EP). The user
    must be aware of the possibilities (or limits)
    of a processor in order to define a "well
    adapted" computation task. This idea is close-
    ly linked to the addressing capacity of an EP
    and more generally to:
- what is the memory hierarchy of the system?
    If the elementary processor is fast, the speed
    with which it will consume and produce the data
    requires a memory with an adapted access time.
    For expense reasons, it is likely that it would
    not be able to contain the entire program (code
    and data). Therefore several memory levels
    have to be taken into account, which brings up
    the next questions:

- how are the data located in the memory? Do they travel the different memory levels?

A standard version of the system would happily dispense the user to explicitly deal with these aspects. However, taking into account the variety of the algorithms used, such a version can never be optimal: for example, access to the point coordinates in a finite difference grid (movement in an array with constant increment) and the same access in a finite element grid (access via a connectivity matrix) cannot be realized within an unique mechanism. A set of primitives in the programming language can help the user to optimize a part of his code: information about data allocation in secondary memory, data movements between primary and secondary memories (and vice-versa), and primitives to operate on sparse data structures.

As regards the synchronization of computation tasks, several theories can be forwarded. An implicit parallelism is made attractive for the user but can prove to be non-optimal in its execution time.

An explicit parallelism between tasks obliges the program to express this parallelism. But in a specific synchronization language, the user will be able to express a maximum of parallelism, as far as well adapted tools are offered like:

- the generic task: a given algorithmic code can be activated on different data. Each task distinguishes itself by a generic index which refers to a block number in a matrix (e.g. a set of lines in the context of blocks iterative scheme or to a number of the sub-domain (nested dissection splitting on a discretized domain)).
- the possibility of exploiting separate sequencing graphs in a pseudo-asynchronismic manner. Only the availability of data can prevent the continuance of these graphs (producer/consumer mechanism during the assembly phase).

2.3. Program Optimization

Finally, we must distinguish between the natural parallelism of the algorithm and the parallelism which the user has to add artificially in order to achieve satisfactory performances. The number of processors in

124

the system must be taken into account. The degree of parallelism must be sufficient to guarantee the activity of a maximum number of processors. This analysis can lead to the adaptation of another splitting of its data structures to increase the number of elementary tasks. Certain methods can, by nature, be well adapted to the number of processors. This is the case with the nested dissections which reveal a variable number of sub-domains depending of the extent of the split.

The latter aspect is undoubtedly the most difficult with which to come to terms. In fact, an optimal program will result from several comparative simulations or executions.

3. ARCHITECTURES MIMD AND PERFORMANCE

To illustrate the MIMD concept, we shall give some possible architectures dedicated to large scale numerical applications. These architectures have been simulated on a benchmark of representative algorithms as presented in Chapter 2. This evaluation has been conducted in the context of the French Supercomputer Project called MARIANNE. Finally, we shall give some simulation results.

3.1. Software Concepts

We shall make a clear distinction between the computation parts of programs and the control part describing concurrency, asynchronism or sequentiality.

A program can be viewed as a collection of "tasks", working on actual objects. A task algorithmically refers to a subroutine, generally existing, executable on conventional processors. The sequencing portion of the program is original since it will express dependences among tasks and thus show concurrency and synchronization.

A closer view of tasks permits to define several kinds:

> (i) A computation task (C-TASK) defines data processing in a number-crunching processor. The user will name actual inputs and outputs handled by a C-task, and subroutines performing the corresponding processing of data.

(ii) A movement task (M-TASK) will re-organize or construct new data objects from others: sparse matrices, subdomain construction are usual in large scale problems. An M-task is intended to run on some dedicated processor capable of handling complex adressing modes and located closer to storage units than number crunching processors.

(iii) A sequencing task (S-TASK) contains rules governing the relative dependences of tasks. The control is explicit and written in a specific language, called XANADU. An S-task will naturally be interpreted in the Main Control Unit of our multiprocessor architecture.

There may be much parallelism between C-tasks or M-tasks. Hence elementary processors should be efficient pipelined units. However, we are much more interested in pure concurrency between tasks, which exhibits an extra parallelism and is the key for higher speed execution over conventional systems.

3.2. Hardware Concepts

The MIMD organization of computers is the single way to overcome technological limits and current architecture bottlenecks in the range of large scale numerical applications. In effect, supercomputers like CRAY-1 or CYBER 205 lead to performances not far from 10 percent of theoretical speed. This somewhat poor result comes from sequential control and from the fact that scalar processing, though fast with the technology used, does not make use of all system resources. The software and hardware arguments developed above give the global hardware characteristics of the possible architectural organizations:

(i) Data processing units (EPs) will be associated with C-tasks. They are assumed identical and may be augmented with cache or local memory modules.

(ii) Data movement units (MPs) are associated with M-tasks. They have no special floating point capabilities, but rather complex addressing units and built-in transfer strategies.

(iii) Memory modules (LMs), visible by data processing and movement units, will contain code and
data currently active.

(iv) Secondary or ternary memory modules (SMs),
supporting the whole code and data bases of
programs.

(v) Depending on architecture types, interconnexion
networks will hook up EPs, MPs, LMs and SMs
together.

(vi) A Supervisor Unit, interpreting S-tasks program,
will coordinate C-tasks and M-tasks execution
according to executable tasks and available
resources.

Two major kinds of MIMD architectures can be raised
from these hypotheses: one with a global main memory
(processors are thus strongly coupled) and one with local
memory modules dedicated to each processor handling one
task completely (processors are loosely coupled). The
relative merits of both approaches can be found in [18].

We now describe the loosely coupled architecture
where several organizations can still be examined.

3.3. Loosely coupled MIMD Architectures

In that context, figure 4 gives three different
organizations, where:
- EPs are pipelined processors in the range of 5
  Mflops performance. Each EP can address either
  one, two or, via a network, any local memory
  modules. There is no direct link between EPs.
- LMs are "local" memories whose features are fast
  access (100ns) and small size (64-256 Kwords).
- MPs are movement processors capable of transferring data/code between SMs and LMs (to form a
  C-task into one LM), or transferring data from
  SMs to others (to execute M-tasks).
- SMs are secondary memories with a large capacity
  (10 to 100 Mwords) and low transfer rates
  (3Mwords/second).
- Interconnexion networks are of Omega-type, asynchronous and adapted to MIMD data routing.

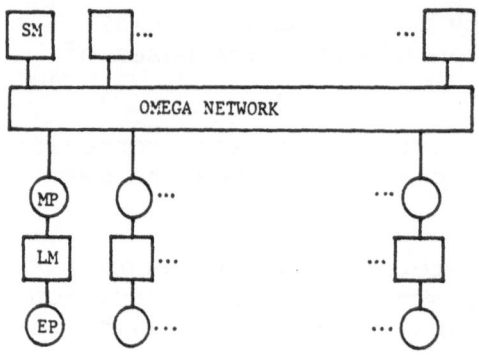

architecture 1

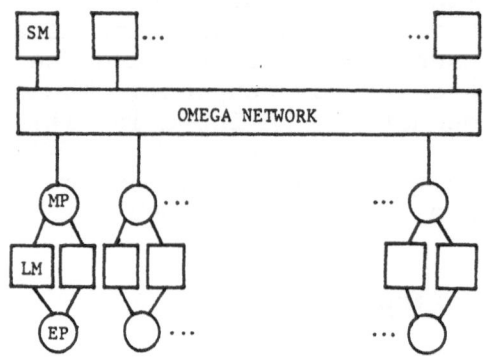

architecture 2

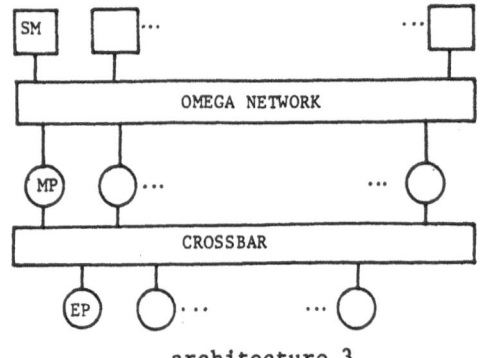

SM : Secondary Memory

MP : Movement Processor

EP : Elementary Processor

LM : Local Memory

architecture 3

Fig. 4. MIMD multi-processor architectures.

The sequencing part of a program, defined in S-tasks, will be interpreted by the Supervisor Unit. Having detected an executable C-task, it will activate different local functions:

1- Make reservation for one LM,
2- Make reservation for the corresponding MP and initiate transfers from SMs to LM: the code and input data are loaded into the LM,
3- Reservation of one EP,
4- De-allocation of MP when inputs are loaded,
5- Run the task code on EP,
6- Wait for "end of execution", then
7- Reservation of MP associated with the LM to unload,
8- Transfer of output data into SMs, and de-allocation of EP,
9- De-allocation of MP and LM.

Architecture 1 is rigid in that it prevents any overlay between task loading phase and task execution. Architecture 2 can deal with that (task A in execution, while task B being unloaded). Architecture 3 gives still more flexibility, but at higher cost, since an additional network is necessary.

Other possible schemes for activating tasks have been studied to avoid unnecessary transfers of data through the memory hierarchy, and improve overall system performance by either keeping data in LMs and execute different codes, or by keeping code in EP and transferring different data blocks.

3.4. Task Sequencing Language

As mentioned, we shall make a distinction between C-tasks (computation), M-tasks (data moves on secondary modules), and S-tasks.

3.4.1. Different task types. A task normally describes all actual object names it manipulates. However, their size can be specified at run time to give the user some flexibility in cases where one task acts on very large data with the same algorithm (just like in a SIMD way). So a task can be defined:

- Normal: all data attributes are known at compile-time
- Generic: data are incompletely specified (in size, not in name). Some parameters under Supervisor

control will be passed to a generic task to complete it. There will be generation of several tasks with same code but different data blocks, executed independently on several processors.

3.4.2. DECLARE and ALLOCATE parts. The DECLARE part contains usual declarations, and some variables specifically managed by the supervisor, the SCHED variables. A SCHED can be declared of either scalar or list type. According to its role in the sequencing, a SCHED supports some information useful for the tasks (generic tasks in particular) or for the supervisor (for taking decisions and tests). A SCHED is generally devoted to handle any control information shared between tasks and their sequencing part.

The ALLOCATE part contains directives for data implementation in secondary memory modules. A large data structure can be split into subsets regularly stored in modules according to the following SPLIT statement:

SPLIT BETA INTO BLOCKS (32,=)
FROM n1 to n2 STEP d START n3;

BETA is decomposed into packets of 32 lines, which are installed into secondary memory banks from n1 to n2 with a step d, packet 0 being in bank n3. This SPLIT statement happens to be often useful, and sometimes mandatory if the user decides to optimize execution by taking care of data transfers and traffic between secondary memory and processors.

3.4.3. Sequencing expressions. Used in S-tasks, the language must have some capabilities to express:

- sequential execution of tasks (or task phases),
- conditional execution,
- iterative execution,
- parallel execution of generic tasks,
- independent execution of different tasks,
- synchronized execution of different tasks.

A possible approach consists in considering the sequencing code as a set of "formulas", non sequentially ordered, describing under which conditions a given set of actions can be "fired" throughout the system. A formula does not have semantic attached to its location in the sequencing, hence it is purely asynchronous with any other one, and obeys the following syntax:

< CONDITIONS > => < ACTIONS > ;

This language called "XANADU" is briefly described in [16] and in more detail in [17].

Another possible approach has been implemented on a subset of the architecture presented here, located at our ONERA company, the system is built around four AP120B array processors, controlled by a supervisor running on a SEL32 minicomputer. Detailed information will be given in the next chapter.

3.5. Simulation Results

3.5.1. Characteristics of the simulator. Designing a high performance system specialized for numerical applications demands evaluation tools capable of showing real evaluation on full scale and real problems of some significance. We kept it in mind when defining our MIMD architecture simulator:

(i) it interprets programs with their actual complexity in size, code and control strategies,

(ii) it makes use of the XANADU language for the control part, and of the number of flops corresponding to each task,

(iii) it will execute the program on a realistic behavior. No modellization in any part of the architecture is used to simulate any component. In particular, all conflict accesses to memories or network are exactly taken into account.

In the context of a loosely coupled system, the simulator is capable of describing several architectures which can be parametrized:

(i) elementary processors (1 to 16, cycle time for a floating point operation),

(ii) local memories (1 to 32, access time),

(iii) movement processors (1 to 16, cycle time),

(iv) secondary memory modules (size is irrelevant, cycle time),

(v) supervisor's cycle time,

(vi) set-up time and routing time in a path of the Omega network.

3.5.2. Performances. The table below gives the performances obtained on a benchmark of programs presented in [18]. These results are based on the following characteristics of the architecture:

- architecture number 2,
- 16 EP of 5 Mflops (effective) per EP,
- cycle time of secondary memory: 300 ns,
- cycle time of local memory: 100 ns,
- length of LM: 64Kwords.

program	performances
BRAIL2	20.8
BRAIL3	14.5
PASFR1	43.8
PASFR2	38.6
PASFR3	38
HOLST	48.7

The most interesting aspect to note is that the performances increase between monoprocessor version (5 Mflops) and multiprocessor one (about 40 Mflops with 16 EP). This 0.5 yield is very attractive to overcome the impressive present and future needs in scientific computation.

4. IMPLEMENTING MIMD CONCEPTS

The hardware and software concepts of loosely coupled MIMD architecture that are described above have been particularly studied in the context of a multi-array-processor configuration. Specific software tools have been developed to allow numerician users to implement parallel applications. Programming and testing real parallel applications shed some light on the problems encountered when expressing and exploiting MIMD parallelism. Performance evaluations allow the user to compare different ways of parallelization and points out the bottlenecks of the configuration. Architecture improvements have been evaluated.

4.1. Multi-Array-Processor Configuration

A Multi-Array-Processor configuration presents interesting features:

- Programming context on pipeline-type array-
  processors is used more and more by numeric-

ians interested in cost/performance ratio.
The mathematical libraries provided on such
machines are improving in connection with user
needs.
- Designing a multiprocessor configuration allows
  modular addition of array processors. System
  growth may be conducted in relation with
  applications size.

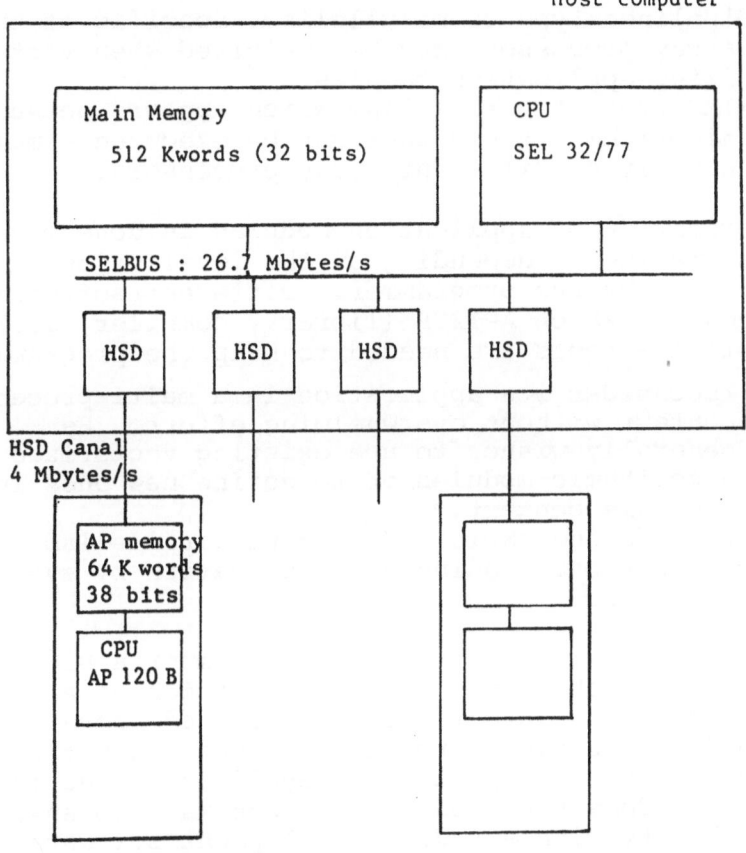

Fig. 5. Multi-array-processor architecture.

The multi-array-processor configuration which has been installed at the ONERA is shown in Figure 5. Three (four in the near future) processors AP120B manufactured by FPS are connected to a mini computer GOULD SEL 32-77 [19].

Such a configuration may be described as a particular loosely coupled MIMD architecture. The host computer acts as a supervisor unit. Each AP120B is a processing element which has its own local memory. The secondary memory, sharable by the processing elements, is the main memory of the SEL computer.

4.2. Parallel language implementation

On such a multi-array-processor architecture, two kinds of parallelism must be taken into account:
- a pipelined type of parallelism, specific of the array processor, may be exploited when vectorizing application modules.
- a MIMD type of parallelism which appears between algorithmic tasks that can be executed simultaneously on different array processors.

Vectorization of application modules is done more or less automatically depending on the level of optimization required by the programmer. Different softwares are provided by FPS on AP120B (library, compiler, ...). But new software tools are needed to help the programmer:
- to reconsider its application in a multi-processor context without overwhelming effort. He generally wishes to use existing vectorized algorithmic modules or to define new ones in the same context.
- to express and exploit MIMD type parallelism without overloading his application by system problems.

The objective is to offer the programmer a high level language which allows the decomposition of a parallel application into tasks as described above. The basic assumption is always to maintain the separation between the algorithmic part of an application and its control part. Some hypotheses have been made to ensure the compatibility with standard development software and existing user's modules. Other assumptions deal with implementation constraints.

The language called LESTAP (Language for Expression and Synchronization of Tasks of a Parallel Application) defines an application as a set of computation tasks and one sequencing task.

- A computation task corresponds to an algorithmic module which has been developed in the host mono-array processor context. Part of the code running on the host computer is written in host FORTRAN, it includes calls to routines executable on <u>one</u> array processor. These routines may have been written in AP-FORTRAN or in APAL assembly language or in VFC (Vector Function Chainer). Data movements between host and an array processor are automatically generated in FORTRAN HASI interface context, but have to be expressed in VFC or APAL.
- The synchronization task is a routine which runs on the host computer and which is in charge of activating the different computation tasks. The synchronization functions provided are described in Figure 6.

The implementation of LESTAP language has been done in the context of the host multitasking system. A system task is associated to each computation task. The system task corresponding to the sequencing task manages the activation of the other system tasks according to the synchronization graph of the application. A precompiler has been implemented: it generates FORTRAN 77 code including calls to functions of a synchronization kernel which interfaces the system primitives level [20][21].

4.3. Programming Parallel Applications

Different applications have been developed in the LESTAP software tools context. Most of them run on two array processors and had been developed on SEL/monoAP configuration. The performance increase obtained is between 1.5 and 1.8 [22][23]. A turbulence numerical simulation (64x64x64 cube) code leads to a 2.3 Megaflops performance. Such a program is intensively exploited at ONERA by physicists concerned in turbulence simulation [24]. Only one application (PDE solving by the sub-domains method) has been implemented on three AP configurations. The performance ratio between one and three processors is 2.5. New parallel applications are under development [25].

- **Sequentiality :**
  C-task T2 must be executed after
  the completion of C-task T1

- **Parallelism** of different tasks
  The n C-task T1,T2,...Tn
  may run in parallel

- **Parallelism** of generic tasks
  Several copies of the same
  C-task T1 may run in parallel
  (on different data)

- **Mixed parallelism**
  Several copies of a generic
  C-task T1 may run in
  parallel with other C-tasks
  T2,...,Tn

- **Iteration**
  C-task T1 will be run
  n times successively

- **Conditional processing**
  According to the value of
  a boolean variable V
  C-task T1 or C-task T2
  will be run

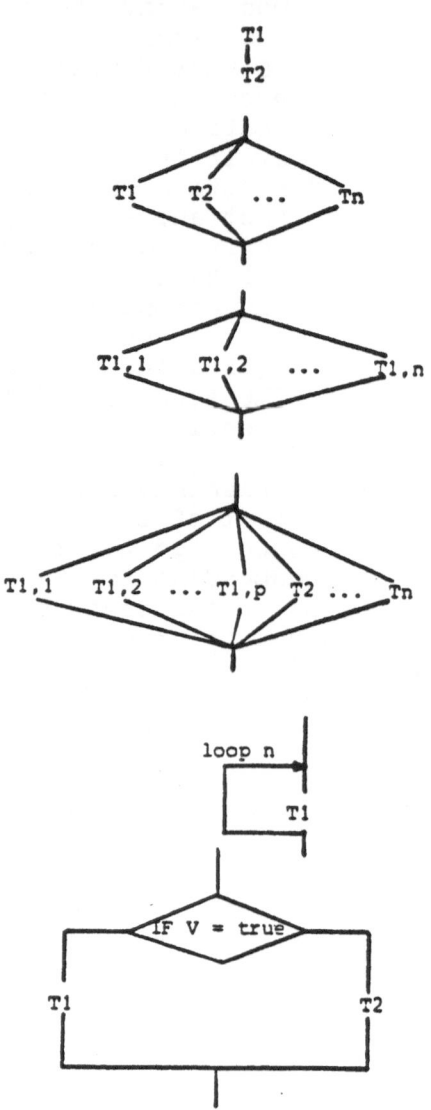

Fig. 6. Synchronization functions of LESTAP language.

136

From the user's point of view, the main problem is how to reach an optimum efficiency. In fact, efficiency obtained is strongly related to the degree of exploitable parallelism provided by the application. But it also depends on the size of data transfers needed across memory hierarchy (disk, host memory, AP memory) compared to computational size. On such a multi-array-processor configuration, most of the applications are often I/O bounded [26]. An optimization of a parallel application is always needed to test:

- different ways of exploiting the parallelism of the application,
- optimal granularity of task according to data transfer time/computing time ratio,
- possible overlap between computation and I/O,
- well adapted data structures.

The LESTAP language is well suited to compare different synchronization schemes. Due to the separation between algorithmic part and control part, only the S-task needs to be modified. But such an optimization is often necessary before coding the computation task. A methodology called the "Program Frame Approach" has been proposed: it mainly consists in executing the control part and transfer operations but in simulating the computation part. After evaluation of the time consumed by the execution of a C-task (operation count), it is encoded by a simple arithmetic operation. Only the transfer operations are exactly coded. This method was used to choose among two sequences of the numerical simulation of turbulence code [27].

4.4. Configuration Evolution: Performance Evaluation

Performances obtained on existing multi-array-processor configurations show how time consuming the data transfers are. The main bottleneck is due to host memory size which is not large enough to anticipate I/O transfers from the disks. For the turbulence simulation program, the performance obtained would have been 3.3 Megaflops (compared to 2.3) on two APs if all the data could have been stored in host memory. This problem of feeding the array processors with data becomes more critical when the number of processors used increases.

The multi-array-processor architecture evolution deals with the adjunction of the fourth processor and the connection of a sharable memory. This additional

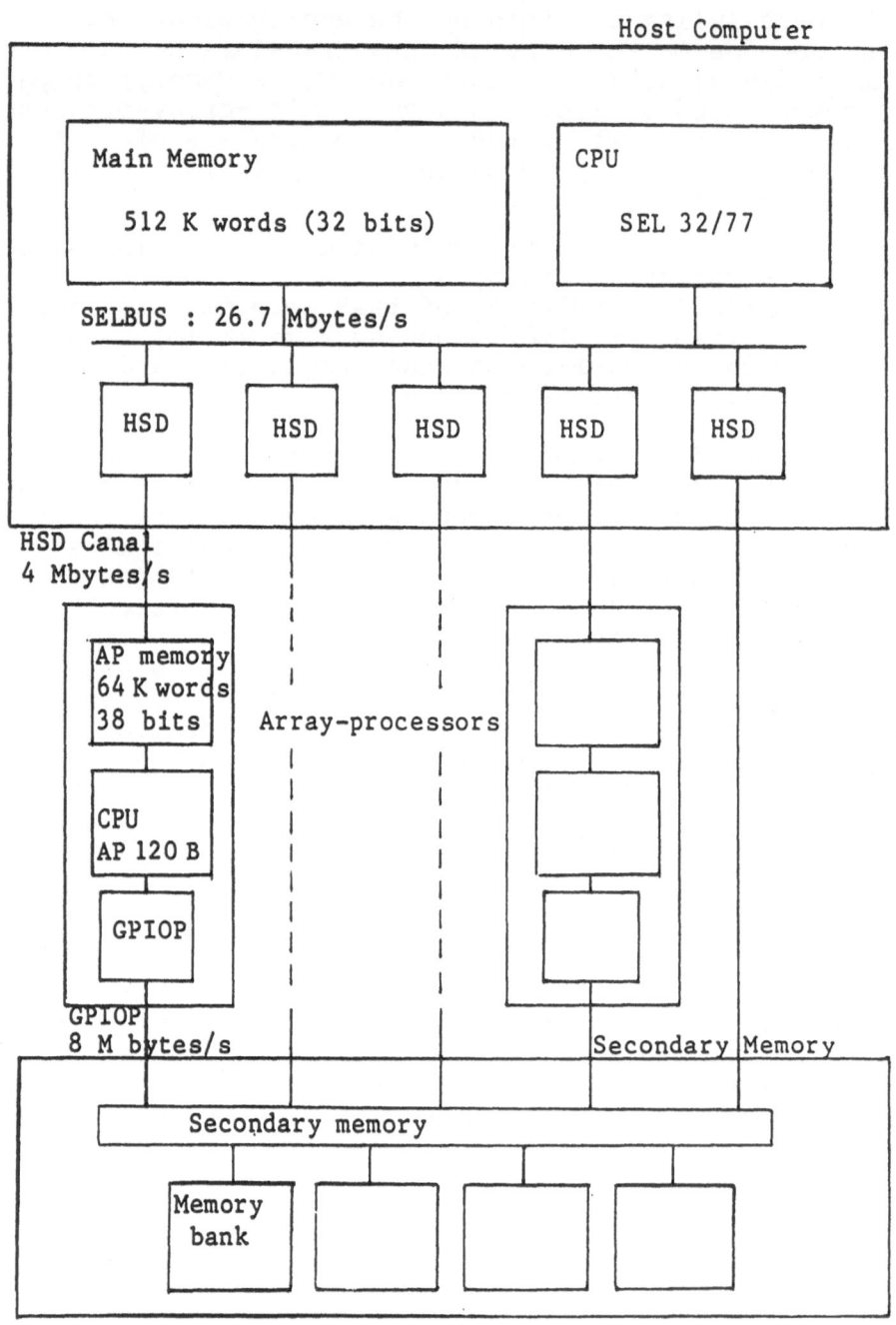

Fig. 7. Multi-array-processor architecture evolution.

memory level will be directly connected to the AP as shown in Figure 7. This architecture evolution leads to a new MIMD configuration where the additional memory constitutes the secondary memory. Each of its four memory banks may be accessed directly or by daisy chaining from each of the AP. This hardware evolution has been evaluated by use of the MIMD loosely coupled architecture simulator which has been described in the previous chapter. This simulator has been parametrized according to SEL/multiAP architecture characteristics (processor speed, memory throughput, host system overhead, supervision cost, ...). A simulation program corresponding to the turbulence numerical simulation code has been developed. It integrates computation task times which have been measured on the configuration. The performance obtained, after memory adjunction, on this program is 4.2 Megaflops (compared to 2.3 on the actual configuration) for two AP. But the same program could run on larger domain (128×128×128 cube) using the four processors. The estimated performance is 8.1 Megaflops [28].

CONCLUSION

While profiting from vectorial parallelism, the objective of the MIMD approach is to increase performances by exploiting parallelism of true multi-processor architectures. Near future supercomputers (CRAY X-MP, CRAY 2) allow the user to express synchronization between tasks constituting a parallel program running on different processing units. In the context of numerical applications, expressing and exploiting such an MIMD parallelism is a major problem.

Language and software tools have been studied in the context of MIMD loosely coupled architectures. A language and a simulator capable of interpreting object codes delivered by a compiler have been designed. Three aspects may be taken into account when using such tools:

- evaluation of new parallel algorithms,
- comparison of different MIMD architectures,
- development of scheduling strategies at the supervisor unit level.

The design of specific tools in the context of a multi-array-processor architecture is an important validation phase of our study. Moreover it allows numerician users to develop and test parallel applications.

ACKNOWLEDGMENTS

This study was done in the context of the French Supercomputer Project called MARIANNE and supported by DRET Contracts.

The ONERA-CERT computer architecture group working on this project is composed of:
M. ADELANTADO, Ph. BERGER, D. COMTE, P. CROS, S. DONNET, Ch. FRABOUL, N. HIFDI, P. SIRON.

REFERENCES

[1] D. Comte, J.C. Syre, Les supercalculateurs, Article de la Recherche, vol. 14, n°147 (sept. 1983).
[2] P.D. Lax, Report of the Panel on Large Scale Computing in Science and Engineering, NSF, Washington (Dec. 26, 1982).
[3] A major international conference on supercomputers, April 1984.
[4] D. Kuck, Structure of computers and computation, Vol. 1, J. Wiley & Sons.
[5] Ph. Treleaven, Future computers: logics, data flow, ... control flow?, Computer (March 1984).
[6] D. Comte, N. Hifdi, J.C. Syre, The Data driven LAU multiprocessor system: results and perspectives, IFIP, Melbourne (1980).
[7] C. Rechbach, Numerical calculation of three-dimensional unsteady flows with vortex sheets, Congrés AIAA sur les Sciences Aéronautiques, Huntville, U.S.A., (January 1978).
[8] S.G. Hotovy, L.J. Dickson, Evaluation of a vectorizable 2D transonic finite difference algorithm, 17th Aerospace Sciences Meeting, New Orleans (January 1979).
[9] Ph. Berger, P. Brouaye, J.C. Syre, A mesh coloring method for efficient MIMD processing in finite element problems, Int. Conf. on Parallel Processing, Bellaire, Michigan (August 1982).
[10] A. George, W.G. Poole, R.G. Voigt, Incomplete nested dissection for solving n by n grid problems. SIAM Num. Analysis, Vol. 15, n°4, (August 1978).
[11] A. Lichnewski, Sur la résolution de systèmes lineaires issus de la méthode des éléments finis par une machine multiprocesseur, Rapport Technique INRIA, n°119 (February 1982).
[12] G.M. Baudet, Asynchronous iterative methods for multi-processors, Rapport Technique Carnegie-Mellon, (1976).

[13] J.C. Miellou, Algorithmes de relaxation chaotique à retards, RAIRO, R1 (1975).

[14] Ph. Berger, Etude, expression et évaluation du parallelisme dans la résolution d'équations aux dérivées partielles sur une architecture MIMD à mémoires locales, Thése Doc.-Ing., ENSAE (December 1982).

[15] C.P. Arnold, M.I. Parr, M.B. Dewe, An efficient parallel algorithm for the solution of large sparse linear matrix equations, IEEE computers, Vol. C-32, n°3, (March 1983).

[16] M. Adelantado, D. Comte, P. Siron, J.C. Syre, A MIMD supercomputer system for large scale numerical applications, IFIP, Paris (1983).

[17] M. Adelantado, D. Comte, S. Donnet, P. Siron, Description du langage XANADU, Rapport final n°3/3210/DERI, Onera-Cert, 2 avenue E. Belin, 31055 Toulouse Cedex, France.

[18] Ph. Berger, Parallélisation d'algorithmes numériques, Rapports n°1/3191/DERI et n°3/3191/DERI, Onera-Cert, Toulouse.

[19] J.P. Boisseau, M. Enselme, D. Guiraud, P. Leca, Potentiality assessment of a parallel architecture for the solution of partial differential equations, La Recherche Aérospatiale 1982-1.

[20] Ch. Fraboul, N. Hifdi, LESTAP: a language for expressing and synchronization of task on a multi-array-processor, Minipaper 1983 FPS Users Meeting, Monterey, Canada (April 24-27, 1983).

[21] Ch. Fraboul, N. Hifdi, LESTAP: Expression et mise en oeuvre d'applications paralléles, Rapport n°1/3192, Onera-Cert, Toulouse.

[22] Ch. Fraboul, D. Guiraud, N. Hifdi, Implementing parallel applications on a multi-array-processor system, Proc. 1st Int. Symp. on a Vector and Parallel Computing in Scientific Applications, Paris (March 1983).

[23] J.P. Boisseau, A. Cosnuau, M. Enselme, Ch. Fraboul, D. Guiraud, N. Hifdi, Le langage LESTAP et son application à la parallélisation d'applications d'aérodynamique, Rapport n°11/3479CY Onera.

[24] P. Leca, Ph. Roy, Numerical simulation of turbulence on minisystems with attached processors, La Recherche Aérospatiale 1983-4.

[25] F. Bourdel, Pré-étude algorithmique de la méthode des sous-domaines de Schwarz, Rapport n°1/3612 Onera-Cert GAN.

[26] G.S. Patterson, Large scale scientific computing: future directions, Comp. Phys. Com. 26 (1982).

[27] M. Enselme, Ch. Fraboul, P. Leca, A MIMD architecture system for PDE numerical simulation, 5th IMACS Int. Symp. on computer methods for partial differential equations, Bethlehem, Pennsylvania, U.S.A., (June 19-22, 1984).

[28] Ch. Fraboul, Evaluation d'évolutions matérielles et logicielles du système parallèle SEL32/multiAP120B, Rapport n°2/3215, Onera-Cert.

IV. APPLICATIONS

SUPERCOMPUTERS IN THEORETICAL HIGH-ENERGY PHYSICS

D. Barkai

Control Data Corporation at the
Institute for Computational Studies at CSU
P.O. Box 1852, Fort Collins, Colorado 80522

K.J.M. Moriarty

Institute for Computational Studies, Department
of Mathematics, Statistics and Computing Science
Dalhousie University, Halifax, Nova Scotia
B3H 4H8, Canada

C. Rebbi

Department of Physics, Brookhaven National
Laboratory, Upton, New York 11973

ABSTRACT

The use of supercomputers to solve fundamental
problems in high-energy physics, e.g. the numerical
solution to Quantum Chromodynamics, is discussed.

INTRODUCTION

It is very good that we are having this meeting on
the use of supercomputers in theoretical science thus
allowing workers in diverse areas of science to share
their experiences and possibly learn from one another.
This cross fertilization between fields has proved particularly fruitful recently in high-energy physics, where
tested methods of condensed matter physics have been
applied successfully. We will discuss this further in
the text.

In section 2 we introduce a short outline of the development of computational physics showing how the growth of computational physics has closely paralled the growth in the power of computers. Such power has evolved to the point that computational physics is now becoming recognized as a field separate from theoretical and experimental physics. Section 3 contains a brief discussion of one of the supercomputers currently available today - the CDC CYBER 205 - and discusses two hardware instructions on this machine that make it particularly suited for Monte-Carlo calculations. We conclude in section 4 with a brief review of a recent study of the force between static quarks using a CYBER 205. This is a very computationally intensive problem which relies heavily on the hardware instructions described in section 3.

COMPUTATIONAL PHYSICS

A supercomputer today may be loosely defined as a machine capable of a computational speed in excess of 100 Mflops. For some typical current machines the following peak rates are achievable in special circumstances: CRAY-1S (160 Mflops), CRAY X-MP-4 (840 Mflops), CDC CYBER 205 (800 Mflops) and Fujitsu VP200 (533 Mflops), which are in the region of two orders of magnitude faster than the fastest serial machines currently available. All of these are examples of supercomputers which achieve their speed by a form of parallelism known as pipelining, which is analogous to processing arithmetic like an assembly line in a factory. Such supercomputers are known as vector processors because the pipelining allows one to calculate with one-dimensional arrays, called vectors, very efficiently. The user has placed at his disposal a tremendous computational power. Projections for the future indicate that we will see even more impressive increases in speed with 10-60 gigaflop machines becoming available soon.

There is a need for supercomputers for scientific and engineering research. Relevant research cannot be carried out without supercomputers in such areas as:

- Fluid dynamics - for full-body analysis of aircraft and the study of turbulence,

- Economic modelling - for modelling the world economy,

- Meteorology - to make accurate weather predictions as often as are needed,

- Oceanography - for three-dimensional tide simulation and the ocean-atmosphere interface studies,

- Quantum chemistry - for the study of large molecules,

- High-energy physics - to analyze the quantum field theory for strong interactions,

- VLSI design - the design of the new generation of chips,

- Computer design - the simulation of the next generation of supercomputers and

- Automobile design - crash analysis.

Although there are over 100 supercomputers in active use outside of secret sites there is <u>very limited</u> access to them for the mass of scientists and engineers so these machines are not having the effect they should. The access to these machines must be increased because:

- the advancement of scientific and engineering research requires state-of-the-art computers and

- the design of products for the market place in as short a time as possible requires them.

In addition to the above, supercomputers should be available, on a limited scale, to university students in order to familiarize them with the new technology. Indeed, training students on serial machines does not prepare them for software design on a vector processor.

Two questions immediately come to mind when discussing vector processors. They are:

1) Are many problems capable of vectorization? and

2) Is vectorization easy and painless?

The answer to both of these questions is yes. Unfortunately, because many workers find vector processors intimidating, many of the codes running on supercomputers today are scalar codes with a minimum of effort put into making them run on vector processors. Many workers feel that obtaining an execution increase of 4 or 5 times that of a CDC 7600 is an adequate mark of success and no further effort need be expended. A proper redesign of the algorithm can result in increases in speed of 25 to 30 times that of the CDC 7600.

In theoretical high-energy physics there has been a dramatic change in the way computers are used. Previously, computers were used merely to supplement analytic calculations, e.g. to calculate some integrals, to obtain some leading eigenvalues, to invert a matrix, etc. However, a large class of researchers would even try to avoid this small use of computers by, for instance, putting a bound on an integral, etc. Now because of the awesome power of the current supercomputers and because no analytic methods exist to solve certain classes of problems, theoreticians simulate a complete system with predetermined dynamics on a supercomputer. So even though we cannot write down results in closed analytic form we can query the computer and obtain numerical answer to our questions.

This development has led to the birth of a new area of physics - called <u>computational physics</u>. (And, of course, the related fields of computational chemistry, computational biology, etc. have also developed). Some workers feel that the development is so significant that they claim physics now consists of the three fields: theory, experiment and computation. To actively pursue computational physics requires that the researcher knows such areas of computing science as computer architecture, operating systems and programming languages, in addition to numerical analysis and, of course, areas of pure and applied mathematics and physics. Thus, the background knowledge of a computational physicist is quite different from that of a traditional theoretician. Computational physics has techniques and methods which are applicable to many diverse areas of physics. We can study common problems in different subjects simultaneously. It is therefore important for computational physicists interested in different areas of research to talk to one another and hold joint meetings such as the present one. As an example of this common interest, it can be stated that the techniques developed in condensed matter physics have been taken over and used to solve problems in high-energy physics.

Of course, this raises many objections from our more traditionally-minded colleagues, such as:

- Is computational physics respectable?

- If one cannot do the problem analytically but can only do it computationally

  a) the problem is not worth doing,

  b) one is not really smart enough and should do something which is less intellectually challenging, or

  c) one is not trying hard enough.

A good example of the use of computers in physics is the study of phase transitions in condensed matter physics. The Ising model is a model for magnetization in solids. Such a model has a critical temperature at which magnetization starts. Near this critical temperature, order parameters such as the magnetization, behave in a power-law manner with critical exponents. The problem for theoretical physics is to calculate these critical exponents. Lars Onsager was able to obtain an exact analytic solution to the two-dimensional Ising model. However, up to now it has not proved possible to generalize the method to the three-dimensional Ising model. K.G. Wilson of Cornell University introduced a method called the renormalization-group method [1], which can be most efficiently implemented on a computer, whereby one simultaneously considers several different lengths of scale to solve the problem and calculate the critical exponents. As a result of this work, computational physics is now much more respectable than it used to be. However, it is still difficult to publish too much technical detail on computational methods in the traditional mainline physics journals. These methods have to be published in specialist computational journals. This seems odd since the computational methods are every bit as valid a solution technique as any analytic method. Other examples of the use of computers to solve basic problems in physics can be found in Ref. 2.

At this point we are faced with two divergent views:

- computational methods should only be respectable
    in so far as they lead to more insight into
    the problem and eventually to the development
    of new analytic methods or

- computational methods can be an end in themselves
    as long as they allow us to answer all the
    questions about a system we wish to ask

The argument between these views is unlikely to be resolved.

An added advantage of computational methods is that they can be used to study systems which would be difficult, dangerous or impossible to study in the laboratory. This would include studying toxic chemicals, chemicals difficult to synthesize and, of course, systems in extreme circumstances, e.g. high temperature or high pressure.

The famine in the availability of supercomputers to the general scientific and engineering community [3] has led a number of university research workers to try to build their own parallel processors. The general approach is to get a superminicomputer such as a DEC VAX 11/780 and use it as a host. Obtain a large number of microprocessors, such as the Motorola 68000 and wire these up in some parallel array. Many claim that this will result in a performance superior to that of the generic supercomputers of today. There are serious difficulties with this approach:

- the development of special purpose processors is very demanding on the scientist's time, much more than elaborating computer codes. Also, the interval between the moment one decides to perform a definite computation and the moment in which results become available is bound to increase, thus slowing the pace of research.

- the risk and consequences of duplication of efforts are more severe.

- theoretical methods (or interests) change rapidly and it is not convenient to have algorithms hardwired and difficult to change.

- the performance and savings realized on a special purpose machine are not always superior to what one can achieve with a general mainframe, especially if the man-hours spent in development are taken into account.

Nevertheless, it is quite possible that on a very specific problem a computer built ad hoc may achieve better performance than a general purpose processor. Even in this case, however, the availability of the former does not imply that the latter becomes useless. Quite to the contrary, the special purpose processor will generate very large amounts of raw data (like a large scale physics experiment) and the speed and flexibility of a powerful, general purpose computer will be ideal to extract from them the maximum amount of significant information. In other words, we envisage a situation in which the special purpose processor, when it can really justify its use, will not be a competitor to the general purpose machine, but rather its complement, both devoted to satisfy the inexhaustible need for computational power of modern research.

In the above there is one area of computational physics we have neglected to mention, namely automated algebra [4]. Up to now in our discussion, computational physics has been interpreted as strictly numerical processing. These numerical computations can be supplemented by algebraic manipulation and, in fact, we can test the efficacy of our numerical methods by testing them against computer derived algebraic results. There are a number of programs now available for algebraic computations:

- REDUCE [5] written by A. Hern,

- SCHOONSCHIP [6] written by M. Veltman,

- MACSYMA [7] written at MIT,

- SMP [8] written at CALTEC and,

- AMP [9] written by J-M. Drouffe.

In general, these programs manipulate strings of characters after we have provided a set of algebraic rules. These programs do nothing that the human worker cannot do but they do it much faster and more accurately. Since it is now accepted that it is a waste of the worker's time to do repetitive numerical calculations by hand, the same attitude is developing in regard to tedious algebraic calculations. An example would be Fourier analysis on a group which is needed in lattice gauge theory calculations. For the group U(1) we have the usual complex Fourier series. The groups of interest are much more complicated than this, e.g. SU(3) with the Fourier series taking the characters of the group as basis states. To calculate such expansions by hand to high order (e.g. 16th order) can take the better part of a year while taking only a few minutes by computer. Much more work in the area of automated algebra needs to be done. Examples of such calculations can be found in Ref. 10.

THE CDC CYBER 205

The CDC CYBER 205 [11] consists of a scalar and a vector unit both of which have access to the main memory. Its large main memory consists of 2, 4, 8 or 16 million 64-bit words. The CYBER 205 is the only supercomputer currently available which supports virtual memory, now 2.2 trillion 64-bit words, which allows one to look at problems far bigger than will fit into main memory. The scalar and vector units may operate in parallel. The scalar unit is of the traditional register-to-register construction while the vector unit is of memory-to-memory construction. Because 16 bits are used to specify the length of a vector, the largest vector which can be processed is $2^{16}-1$ or 65,535. The CYBER 205 is available with 1, 2 or 4 vector pipelines, with all pipes performing the same instruction and each pipe producing a result every 20 nsec clock period. The CYBER 205 also supports another unusual feature - 32-bit arithmetic. When operating in the 32-bit mode, the effective number of pipelines is doubled, thus doubling the arithmetic processing rate.

On the CYBER 205 a vector is an ordered set of elements contiguously located in main memory with unit stride. This differs from other supercomputers which allow positive and negative constant stride but support much shorter vectors. The elements of a vector may be 1-bit, 8-bit bytes, 32-bit "half-words" or 64-bit words.

In a book [12] by Iverson in 1962 it was suggested that Fortran did not cater to the needs of scientists and engineers who wished to process mathematical structures such as vectors or matrices. His suggestion led to the development of APL-A Programming Language. It also led to the development of the idea within Control Data Corporation that a vector processor should be developed with Iverson's APL instructions hardwired into the machine. This idea was brought to fruition with the introduction of the CDC CYBER 205 in 1981. There are a host of such application primitives in the CYBER 205 instruction set, e.g. COMPRESS, DECOMPRESS, MERGE, EXPAND, GATHER, SCATTER, etc. For our later discussion we will only be interested in the GATHER/SCATTER instructions so we will limit our discussion to these.

The GATHER instruction gets elements from a source vector in main memory guided by an index-list and stores them in a resultant vector in the order in which they have been obtained. As an example, consider the following source vector and input index-list:

SOURCE VECTOR: 20., 18., 16., 14., 12., 10.

INDEX-LIST: 4, 2, 1, 4, 3, 3, 5, 2, 1, 6

which would produce the resultant vector

RESULTANT VECTOR: 14., 18., 20., 14., 16., 16., 12., 18., 20., 10.

This example illustrates several features of the GATHER instruction:

    1) Any source vector element may be output any number of times during the operation, and

      2) the resultant vector has the same length as the
         index-list and may be shorter or longer than
         the source vector.

To calculate the cost of a vector operation we have two
things to consider, the start-up time and the stream
time. The start-up time is independent of the vector
length while the stream time depends on the vector length.
As a result, to make effective use of the CYBER 205 we
should try to minimize the total number of start-ups and,
simultaneously, make the vector lengths as long as
possible. The SCATTER instruction performs the reverse
operation of the GATHER instruction.

THE MONTE-CARLO METHOD FOR LATTICE GAUGE THEORY

     The theory of strong interactions is believed to be
Quantum Chromodynamics, called QCD for short. In this
theory hadronic particles are viewed as being made up
of still more basic building blocks called quarks, with
the quarks interacting via the exchange of gluons. The
quarks are described by three-dimensional complex vectors
and the Lagrangian of the theory has SU(3) symmetry.
The gluons are described by gauge fields. Until
recently it has proved difficult to make detailed
comparisons of the theory with experiment and thus
establish the validity of the theory. Lately, however,
it has become possible to calculate theoretically some
well established physical observables - such as the
masses and the magnetic moments of the elementary particle spectrum. In this section, we will discuss some
preliminary calculations on the road to carrying out such
comparisons. In particular we will discuss some recent
significant results on the force between static quarks.

     QCD is a relativistic quantum field theory and
suffers from divergences which plague all quantum field
theories. These divergences arise from two sources.
The infinite extent of four-dimensional space-time leads
to infrared divergences while the continuity of space-
time leads to ultraviolet divergences. In order to
regulate these divergences Wilson [13] and Polyakov [14]
introduced the idea of replacing space-time by a discrete
lattice of finite extent. By means of a Wick rotation,
in which we analytically continue t to it, Minkowski
space-time becomes Euclidean space-time. By quantizing
the theory by the Feynman path integral approach, the
equivalence with a special classical statistical mechanical system can be established. We can thus use all the

well established methods of condensed matter physics [15] to analyze QCD and measure physically significant observables.

In our regular four-dimensional space-time lattice, called a hypercubic lattice, the discrete points are called sites while the lines joining sites are called links. The SU(3) symmetry is modelled by SU(3) matrices (3×3 unitary-unimodular matrices) attached to the links. These matrices are the fundamental dynamical variables of the gluonic medium and are responsible for the inter-action among quarks. The smallest plane geometric figure that one can form on the lattice is a square, commonly referred to as a plaquette. The action of the theory is constructed by forming the sum of the matrix products along the plaquette boundaries for all possible plaquettes of the lattice.

Our first problem is to decide how big the space-time lattice should be. It must be large enough to accommodate a typical particle such as a proton yet fine grained enough to account for the quark interactions. It has been found that this demands a lattice of at least 10 sites in each space-time direction. To be on the safe side, we have chosen to use a lattice of size 16^3 in the space directions and 32 in the time direction, i.e. $16^3 \times 32$ [16]. For this size lattice, we have $4 \times 16^3 \times 32$ links to consider and hence the same number of SU(3) link matrices. Each SU(3) matrix has 8 independent variables so we have $8 \times 4 \times 16^3 \times 32 = 4.194304$ million degrees of freedom. In practice, it is useful to represent an SU(3) element in full matrix notation, so that we have 9.437184 million real variables to consider.

The integral we need to calculate is related to the partition function and it is an integral over all link matrices in the lattice. Direct integration is not possible so we use importance sampling methods, i.e. Monte-Carlo methods. Thus, in the above approach to calculations with QCD we are engaged in a statistical experiment with the four-dimensional lattice being like a four-dimensional "crystal". By sampling from this crystal we are able to measure the value of observables e.g. the force between static quarks.

The Monte-Carlo method we employ in our calculations is the method of Metropolis et al. [17]. Typically one starts out the experiment with the SU(3) matrices either all ordered (i.e. 3×3 unit matrices) or random (with randomly chosen elements). By the Monte-Carlo method one replaces the matrices already in place by new matrices chosen according to a large table of SU(3) matrices by means of a suitable rejection or acceptance criterion. One continues to carry out this procedure until one reaches what we refer to as an equilibriated gauge field configuration.

The results of the statistical experiment are then obtained by averaging over many different equilibriated configurations. The statistical error in the measurements can be reduced by increasing the size of the sample used in the averages. The generation of one gauge field configuration or one iteration through the lattice results in more than 10^9 arithmetic operations. One needs thousands of iterations for a good measurement.

To update one of the SU(3) link matrices requires all its nearest neighbor SU(3) link matrices in all four directions. With the periodic boundary conditions, it is impossible to order the link matrices in main memory such that they are contiguous and ready for processing. However, with the use of the data motion instruction GATHER the link matrices can be arranged in the correct order and when processed, the matrices can be put back in place with a SCATTER operation. Such operations proceed at rates comparable to that of vector arithmetic. The fact that the GATHER/SCATTER instructions are hardwired on the CDC CYBER 205 makes it very efficient for Monte-Carlo calculations.

Basically the updating procedure consists of multiplying SU(3) matrices. However, working with vectors of length 3 is not making efficient use of the vector processor. We therefore try to update as many independent link variables simultaneously as possible. Since we only have nearest link interactions we are led to the "red-black" or checkerboard ordering of link matrices. As a result, we are able to form vectors of length $\frac{1}{2} \cdot n^3$ where n is the space lattice size. For $n = 16$, this results in vectors of length 2,048. In order to reduce the computational overhead, we make a number of attempts, between 8 and 10, to make an update before beginning the next iteration. This is called making a number of "hits-per-link".

We now wish to discuss the performance of our code, which was run on a 2 Mword 2 vector pipeline CDC CYBER 205 at Fort Collins, Colorado which, when operating in the 32-bit mode, is rated at a maximum speed of 200 Mflops (without linking). All our calculations were carried out in 32-bit arithmetic except for exponentation and random number generation. The peak rate achieved on the machine in our calculations is 182 Mflops with a sustained rate of 130 Mflops. For our calculations, we required 5 times as much memory as normally available on the system. By a proper use of the memory management we are able to achieve 98% CPU utilization (using virtual memory). The parameter which measures the speed of our algorithm is what is referred to as the "link-update-time". The link-update-time was 40.9 μsec, and is to be contrasted with 1.1 msec on the CDC 7600, a high-perform-ance serial machine.

In our calculations we determined the force between static quarks [18]. We found that the Monte-Carlo results are well described by a Coulomb-like behavior, with a logarithmically varying coupling constant, at short distances, as theoretically expected, and a constant part at large distances, a signal for quark confinement.

We have recently extended our calculations to the measurement of particle masses. This requires finding the quark propagators which involves solving a linear system of equations with over 7 million non-zero coefficients. In a preliminary implementation of a conjugate gradient solver [19] we developed, we found that a matrix with over 3/4 million unknowns can be iterated in about 1.3 sec. of CPU time (it is I/O bound due to the large amount of data). This is a quite good performance. This performance is helped by the exist-ence on the CDC CYBER 205 of a hardwired dot product instruction.

ACKNOWLEDGMENT

We would like to thank Control Data Corporation for the award of time on the CDC CYBER 205 at the Institute for Computational Studies at Colorado State University where the codes described in the text were developed.

REFERENCES

[1] K.G. Wilson, Phys. Rev. B4:3184 (1971).
[2] D.R. Hamann, Physics Today 36:25 (1983).
    S. Wolfram, Sci. Am. 251:188 (1984).
[3] Report of the Panel on Large Scale Computing in Science and Engineering, P.D. Lax, Chairman, US Department of Defence and National Science Foundation (1983).

[4] R. Pavelle, M. Rothstein and J. Fitch, *Sci. Am.* 245: 136 (1981).

[5] A.C. Hern, *REDUCE2 User's Manual*, UCP-19 (1974).

[6] M. Velman, *Comput. Phys. Commun.* 3 (Supplement):75 (1972).

[7] *MASCYMA, Reference Manual* (Version Nine), The Math-lab Group, Laboratory for Computer Science, MIT, Boston, Massachusetts, 1977.

[8] S. Wolfram, *SMP Reference Manual*, Inference Corporation, Los Angeles, 1983.

[9] J-M. Drouffe, *Algebraic Manipulation Program* (User's Manual), Version 6, Saclay Report, (1981).

[10] See, for example, J-M. Drouffe and K.J.M. Moriarty, *Phys. Lett.* 105B:449 (1981) for a use of Ref. 9 and K.J.M. Moriarty and S. Samuel, *Phys. Rev.* 27: 982 (1983) for a use of Ref. 7.

[11] *CDC CYBER 200 Model 205 Computer System - Hardware Reference Manual*, No. 60256020, Control Data Corporation, Minneapolis, Minnesota.

[12] K.E. Iverson, *A Programming Language* (John Wiley, New York, 1962).

[13] K.G. Wilson, *Phys. Rev.* D10:2445 (1974).

[14] A.M. Polyakov, unpublished.

[15] K. Binder, Ed., *Monte Carlo Methods in Statistical Physics*, (Springer Verlag, Berlin, 1979).

[16] D. Barkai, K.J.M. Moriarty and C. Rebbi, *Comput. Phys. Commun.* 32:1 (1984).

[17] N. Metropolis, A.W. Rosenbluth, M.N. Rosenbluth, A.H. Teller and E. Teller, *J. Chem. Phys.* 21:1087 (1953).

[18] D. Barkai, K.J.M. Moriarty and C. Rebbi, *Phys. Rev.* D30:1293 (1984); ibid. 30:2201 (1984).

[19] D. Barkai, K.J.M. Moriarty and C. Rebbi, *Comput. Phys. Commun.* 36, 1 (1985).

THE USE OF A VECTOR COMPUTER IN AB-INITIO PHONON CALCULATIONS IN SEMICONDUCTORS

P.E. Van Camp and J.T. Devreese°

University of Antwerp (RUCA)
Groenenborgerlaan 171,
B-2020 Antwerpen, Belgium

I. INTRODUCTION

There are two ab-initio methods for the calculation of the phonon frequencies of solids: 1) the total energy difference method [1], where the energy difference between the distorted and the undistorted crystal is calculated directly, and 2) the dielectric screening method [2].

In the latter method the electrons in the crystal are treated in the framework of linear response theory. The density response of the electrons is caused by a perturbation due to the motion of the ions around their equilibrium positions. The energy difference of the crystal between the distorted and undistorted configurations is expanded to second order in the ionic displacements. The matrix of the expansion coefficients whose eigenvalues yield the phonon frequencies is expressed in terms of the electron density response matrix. The advantage of the dielectric screening method is that it can treat an arbitrary wave vector but it is limited to harmonic phonons. The present authors have used this dielectric screening method for all their ab-initio

---

° Also at: Department of Physics, University of Antwerp (UIA), B-2610 Antwerpen-Wilrijk, Belgium, and University of Technology, Eindhoven, The Netherlands.

Calculations of the macroscopic dielectric constant and the phonon frequencies of Si.

The electron energies and wave functions needed in the density response matrix are calculated in the local density approximation. The crystal pseudopotential in the Hamiltonian is obtained self-consistently from the same electron-ion pseudopotential used in the electronic part of the dynamical matrix. Furthermore the same exchange-correlation effects taken into account in the Hamiltonian have been used to derive the electron density response matrix. Our investigations show that both consistencies, i.e. in the potentials and in the exchange-correlation, are crucial in the dielectric screening method.

In previous work[3] the present authors obtained phonon disperions curves of Si where the polarizability matrix was approximated by a moment expansion. In the present work this matrix is evaluated straightforwardly by means of a direct summation over all the conduction bands available from diagonalization of the Hamiltonian matrix. The lattice parameter is not taken from experiment but instead calculated from minimalizing the total crystal energy with respect to ionic displacements. The convergence of the phonon frequencies in terms of the number of reciprocal lattice vectors used in the Hamiltonian and in the linear response matrices is investigated in detail.

Because the programs to make the above described calculations take a large amount of computer time the existing codes have been adapted to run on a vector computer. They were then executed on a CDC CYBER 205, where the program as a whole ran five times faster than in scalar mode. In some of the matrix handling routines the gain in time was as high as 70.

II. THEORY

The essential quantities of the dielectric screening theory are the density response matrix χ, the polarizability matrix $\tilde{\chi}$ and the dielectric matrix ε.

The total electronic energy of the ground state of the system is expanded in a Taylor series with respect to the ionic displacements. The density response matrix χ is defined by the linear relationship between the externally applied potential δv^{EXT} and the induced charge density $\delta\rho$:

$$\delta\rho = \tilde{\chi} \; \delta V^{EXT} \tag{1}$$

The polarizability matrix $\tilde{\chi}$ is the factor of proportionality between the charge density $\delta\rho$ and the induced potential δV^{IND}:

$$\delta\rho = \tilde{\chi} \; \delta V^{IND} \tag{2}$$

Eqs. (1) and (2) are given in matrix notation, i.e. χ and $\tilde{\chi}$ are matrices, and $\delta\rho$, δV^{EXT} and δV^{IND} are column vectors.

Using first order perturbation theory one can express the polarizability matrix as follows (in reciprocal space) [2]:

$$\tilde{\chi}(\vec{q},\vec{G},\vec{G}') = \frac{1}{\Omega} \sum_{\ell m} \frac{\eta_\ell - \eta_m}{E_\ell - E_m} <\ell | e^{-i(\vec{q}+\vec{G})\vec{r}} | m>$$
$$<m | e^{i(\vec{q}+\vec{G}')\vec{r}} | \ell> \tag{3}$$

Ω is the crystal volume, $\vec{G}$ ($\vec{G}'$) are reciprocal lattice vectors and $\vec{q}$ is the phonon wave vector. It should be noted that the summations in Eq. (3) in principle run over <u>all</u> conduction bands of the crystal.

The expression of the dielectric matrix ε depends on the particular form of the one-electron equations used. In the local density approximation [4], which is used in the present work, the following form of the dielectric matrix ε is obtained [5]:

$$\varepsilon = 1 - v_c \; \tilde{\chi}(1 - v_{xc} \; \tilde{\chi})^{-1} \tag{4}$$

Here v_c is the Fourier transform of the Coulomb potential and v_{xc} is given by:

$$v_{xc}(\vec{q},\vec{G},\vec{G}') = \frac{1}{\Omega} \int d\vec{r} \int d\vec{r}' \; e^{i(\vec{q}+\vec{G})\vec{r}} \frac{\delta \; V_{xc}[\rho]}{\delta\rho}$$
$$e^{-i(\vec{q}+\vec{G}')\vec{r}'} \tag{5}$$

with $V_{xc}[\rho]$ the exchange-correlation energy. In this work the Slater expression for V_{xc} is used i.e. [6]

$$V_{xc}[\rho] = -\alpha \; \frac{3}{2} \; (\frac{3}{\pi} \; \rho)^{1/3} \tag{6}$$

It should be noted that other expressions (e.g. Wigner [7] or Ceperley-Alder [8]) can equally well be employed without any difficulty. In the present work a value of $\alpha = 0.794$ is used. This value brings the Slater expression into agreement with the Wigner expression at the average density of Silicon. The density response matrix χ (Eq. (1)) is related to the inverse dielectric matrix ε^{-1} through:

$$\varepsilon^{-1} = 1 + v_c \chi \tag{7}$$

and is calculated by numerical inversion of ε.

The dynamical matrix of the system determines the frequencies of the ionic vibrations. This matrix is given by:

$$D_{ij}^{en}(\vec{q};ab) = \sum_{GG'} X_{ij}(ab;\vec{q},\vec{G},\vec{G}')$$

$$- \delta_{ab} \sum_c X_{ij}(ac;\vec{q},\vec{G},\vec{G}') \tag{8}$$

with

$$X_{ij}(ab;\vec{q},\vec{G},\vec{G}') = \frac{1}{\Omega_c \sqrt{M_a M_b}} (\vec{q}+\vec{G})_i \times$$

$$V_a(\vec{q}+\vec{G}) \, e^{i\vec{G}\vec{R}_a} \chi(\vec{q},\vec{G},\vec{G}') \, e^{i\vec{G}'\vec{R}_b} V_b(\vec{q}+\vec{G}') (\vec{q}+\vec{G}')_j \tag{9}$$

Here V_a is the Fourier transform of the electron-ion potential of the a-th ion in the unit cell with mass M_a. Ω_c is the unit cell volume.

The eigenvalues of the dynamical matrix are the phonon energies of the crystal. Since Eq. (9) is completely general no restriction on the phonon wave vector $\vec{q}$ appears in this method. The generalized acoustic sum rule [9] is an important internal consistency requirement of the phonon calculation. It is given by the relation:

$$\lim_{\vec{q}\to 0} \sum_{G,a} (\vec{q}+\vec{G}) \chi(\vec{q},\vec{G}',\vec{G}) V_a(\vec{q}+\vec{G}) \, e^{-i\vec{G}\vec{R}_a} =$$

$$\lim_{q\to 0} (\vec{q}+\vec{G}') \rho(\vec{G}') \tag{10}$$

For $\vec{G}'=0$ Eq. (10) can be rewritten as:

$$\frac{4\pi}{M} \lim_{q \to 0} \sum_{G,a} \frac{\vec{q} \cdot (\vec{q}+\vec{G})}{q^2} \chi(\vec{q},\vec{0},\vec{G}') \; V_a(\vec{q}+\vec{G}) \; e^{-i\vec{G}\vec{R}_a} = \omega_p^2$$

$$(11)$$

with ω_p the ionic-plasma frequency and M the ionic mass. Physically this sum rule guarantees that the effective ionic potential is invariant under a uniform translation so that the acoustic modes have zero frequency at vanishing wave vector.

Expressions (5) and (7) have been used to calculate the microscopic (Eq. (4)) and macroscopic (Eq. (7)) dielectric functions of Si [5], Ge [10] and C and α-Sn [11].

III. COMPUTATIONAL PROCEDURES

1. The Electron-Ion Potential

Several forms of the Si ionic pseudopotential can be found in the literature [12],[13]. In the present work we use the Schlüter-Chelikowsky-Louie-Cohen potential [12] given by:

$$V(q) = \frac{4\pi Z V_1}{q^2} (\cos V_2 q + V_3) \; e^{V_4 q^4}$$

$$(12)$$

with (in au)

$$V_1 = 1.5432 \qquad V_3 = -0.3520$$
$$V_2 = 0.7907 \qquad V_4 = -0.01807$$

Strictly speaking V(q) given in Eq. (12) is not an ab-initio potential (it was fitted to experimental values of valence and conduction band energies). Recent tests performed by the present authors using the Topp-Hopfield potential [13] indicate that the overall featuresof the phonon frequencies calculated with both potentials are similar.

2. The Electronic Band Structure

In the local pseudopotential formalism the Kohn-Sham equations are replaced by (in au)

$$(-\frac{1}{2} \nabla^2 + V_p(\vec{r})) \; \psi_{kn}(\vec{r}) = E_{kn} \; \psi_{kn}(\vec{r})$$

$$(13)$$

where V_p is the local pseudopotential. Since both the pseudopotential V_p and the dynamical matrix depend on the electron-ion potential a self-consistent pseudopotential band calculation must be performed. The computational procedure is as follows:

1. Determine the ionic potential.
2. Choose an empirical starting pseudopotential.
3. Diagonalize the Hamiltonian in a plane wave basis. Typically 193 plane waves are used in the present work while additionally some 100 plane waves are treated in second order perturbation theory.
4. Next construct the charge density ρ.
5. Calculate the Hartree and exchange-correlation potentials. Together with the ionic potential the total pseudopotential can then be constructed.

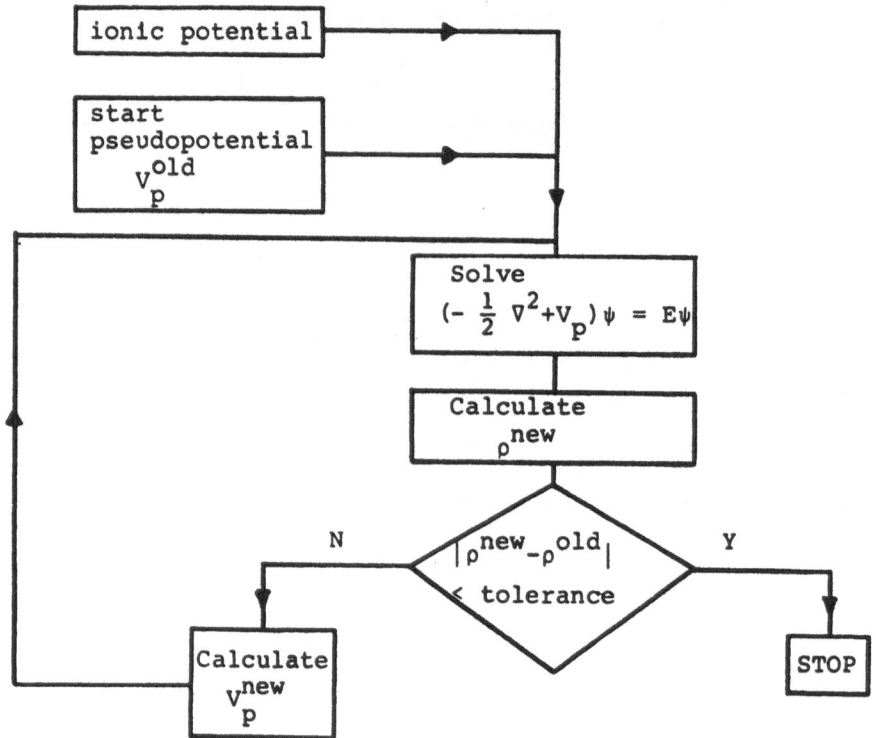

Fig. 1. Flow diagram of the self-consistent band calculation.

6. Replace the starting pseudopotential by the new one and repeat the calculation, i.e. go back to step 4. The iteration cycle is stopped if this new charge density differs by less than a predefined amount from the previous charge density. Convergence is normally reached with less than ten iterations.

In figure 1 a flow diagram of the self-consistent band calculation described above is given.

3. The Total Crystal Energy

In the local density approximation [4] the total crystal energy can be written as

$$E_T = \sum_{kn} E_{kn} - \frac{1}{2} \int d\vec{r} \int d\vec{r}\,' \frac{\rho(\vec{r})\,\rho(\vec{r}\,')}{|\vec{r}-\vec{r}\,'|}$$

$$+ \int d\vec{r}\, \rho(\vec{r})\, [\varepsilon_{xc}(\rho) - V_{xc}(\rho)] \qquad (14)$$

The total energy E_T is minimal at the equilibrium distance of the crystal. Therefore E_T is minimalized with respect to the lattice constant. The value obtained for Silicon is 5.373 Å to be compared with the experimental value of 5.429 Å [14]. This value is in agreement with other calculations using a local pseudo-potential (see e.g. [15]). All the calculations discussed in the next sections were performed using this calculated lattice constant.

4. Vectorization of the Codes

In the phonon calculation, described above, three matrix operations are employed frequently: multiplication, inversion and diagonalization. The dimension of the matrices will be denoted by N.

4.1. Matrix multiplication. A general matrix multiplication can be performed in a variety of ways. Here only the standard algorithm (i.e. the inner product of a row with a column) is used. The number of floating point operations is then $2N^3-N$. It should be noted that there exist algorithms that need only $O(N^{2\cdot8})$ operations [16].

The classical way to program the matrix product $A = B\times C$ is:

```
      DO   10   K = 1,N
        DO   20   I = 1,N
  20  A(I,K) = B(I,1)    C(1,K)
        DO   30   J = 2,N
        DO   40   I = 1,N
  40  A(I,K) = A(I,K) + B(I,J) ⋇ C(J,K)
  30    CONTINUE
10    CONTINUE
```

Timing results of this code executed in scalar mode on a
CYBER-205 are given in table I. In vector mode the
compiler will vectorize loop 20 and loop 40 (which is a
linked triad). Timings are shown in table I both for a
1- and 2-pipe CYBER 205. There are several other possibilities to do the same thing e.g. the use of the special
call "Q8SDOT". However, if we simply unroll loop 40 the
code will run 10 times faster in scalar mode than the
original one: unrolling to a depth of 4 results in the
following code:

```
      DO   10   K = 1,N
      DO   20   I = 1,N
  20  A(I,K) = 0.
      DO   30 J = 4,N,4
      DO   40 I = 1,N
  40  A(I,K) = ((((A(I,K) + C(J-3,K) ⋇ B(I,J-3))
    .    + C(J-2,K) ⋇ B(I,J-2)) + C(J-1,K) ⋇ B(I,J-1))
    .    + C(J,K) ⋇ B(I,J))
  30    CONTINUE
  10    CONTINUE
```

Table I. Time (in sec) for full matrix multiplication as
          a function of the dimension of the matrices.
          The last column gives the time ratio of the
          scalar run over the vector run on a 2-pipe
          CYBER 205.

dimension	scalar	scalar unrolled	vector 1-pipe	vector 2-pipe	ratio
50	0.182	0.023	0.009	0.008	22.8
100	1.457	0.172	0.046	0.037	39.4
150	4.920	0.578	0.126	0.094	52.3
200	11.667	1.353	0.265	0.189	61.7
250	22.793	2.642	0.476	0.326	69.9
300	39.393	4.539	0.775	0.512	76.9

Table II. Time (in sec) for full general matrix inversion
as a function of the dimension of the matrix.
The last column gives the time ratio of the
scalar run over the vector run on a 2-pipe
CYBER 205.

dimension	scalar	vector 1-pipe	vector 2-pipe	ratio
50	0.151	0.010	0.009	16.8
100	1.260	0.050	0.039	32.3
150	5.157	0.136	0.101	51.1
200	14.956	0.282	0.202	74.0
250	34.150	0.509	0.349	97.9
300	67.043	0.825	0.552	121.5

4.2. Matrix inversion. There are several methods
to invert a general full matrix. The one used here is
Jordan's method with partial pivoting. In this method
the given matrix is reduced to a unit matrix by means of
successive elementary operations on the rows. The same
operations performed on a unit matrix gives the inverse
matrix of the original. This algorithm needs $2N^3+N$
floating point operations and is not yet vectorized
completely. In table II timing results are given in
scalar and vector mode on a 1- and 2-pipe CYBER 205.

4.3. Matrix diagonalization. One of the fastest
algorithms to find all the eigenvalues and all or some
of the eigenvectors of a real symmetric matrix is the
Householder QR inverse iteration [17]. This algorithm
consists of three parts:
- Householder transformations to reduce the given matrix
  to a tridiagonal form. This step involves N-2 similar-
  ity transformations
- QR iteration to find the eigenvalues of the tridiagonal
  matrix. In this step the tridiagonal matrix is decomposed into a product of an orthogonal matrix with
  an upper triangular matrix
- inverse iteration to determine the needed corresponding eigenvectors of the tridiagonal matrix. These are
  then backtransformed to yield the eigenvectors of the
  original matrix.
The whole algorithm takes $\frac{5}{3} N^3 + \frac{39}{2} N^2$ floating point

operations (in the case one needs all the eigenvectors).
Unfortunately the inverse iteration step, in its present

Table III. Time (in sec) for matrix diagonalization as
a function of the dimension of the matrix.
The last column gives the time ratio of the
scalar run over the vector run on a 2-pipe
CYBER 205.

dimension	scalar	vector 1-pipe	vector 2-pipe	ratio
50	0.325	0.084	0.083	3.9
100	2.374	0.367	0.348	6.8
150	8.293	0.882	0.838	9.9
200	23.374	1.669	1.578	14.8
250	52.193	2.777	2.583	20.2
300	100.616	4.208	3.948	25.5

form, is not yet vectorized. Furthermore, the Householder
transformation could only be partly vectorized. In table
III timing results are given in scalar and vector mode on
a 1- and 2-pipe CYBER 205. One notices that the ratio
(scalar/vector time) is a factor 3-5 smaller than for
the matrix multiplication and inversion.

4.4. The polarizability matrix. In the pseudopotential method the electron wave functions are expanded
in plane waves:

$$\psi_{kn}(r) = \frac{1}{\sqrt{\Omega}} \sum_G C_k^n(\vec{G}) \; e^{i(\vec{k}+\vec{G}) \cdot \vec{r}} \tag{15}$$

$\vec{k}$ is the electron wave vector (restricted to the first
Brillouin zone and n is the band index. The sum in Eq.
(15) in principle runs over the infinite set of reciprocal
lattice vectors $\vec{G}$, but in practice only a limited number
are taken into account. In the present work up to 193
lattice vectors are included. Inserting the expansion
Eq. (15) into the Kohn-Sham equations leads to an eigenvalue problem:

$$H \; C = E \; C \tag{16}$$

where C is the matrix which contains the expansion coefficients $C_k^n(\vec{G})$ in its columns, H is the plane wave
representation matrix of the Hamiltonian (which is a
real symmetric matrix in the present case) and the
diagonal matrix E consists of the one-electron energies.
Using the expansion Eq. (15) in the expression for the
polarizability matrix Eq. (3) gives:

168

$$\chi(\vec{q},\vec{G},\vec{G}') = \frac{1}{\Omega} \sum_{knn'} \frac{\eta_{kn}-\eta_{k+qn'}}{E_{kn}-E_{k+qn'}} T^{nn'}_{k\ k+q}(\vec{G})\ T^{nn'}_{k\ k+q}(\vec{G}')$$

(17)

with

$$T^{nn'}_{kk'}(\vec{G}) = \sum_{\vec{G}'} C^{n}_{k}(\vec{G}')\ C^{n'}_{k'}(\vec{G}'-\vec{G})$$

(18)

The sum in Eq. (18) resembles a scalar product, the difference being that the products are <u>not</u> taken component by component (except when $\vec{G}=0$). The elements of the second vector ($C^{n'}_{k'}(\vec{G}'-\vec{G})$) are not taken in the same sequence as those of the first vector. In order to be able to use the dot product the following procedure was adopted:

1. Form the difference vector $\vec{G}'-\vec{G}$. Each element of this vector is a three-dimensional vector itself.
2. Check if the difference $\vec{G}'-\vec{G}$ belongs to the reciprocal lattice vector set considered (i.e. one of the first 193 reciprocal lattice vectors). This is easily done using a vector relational and a bit vector. If the difference does not belong to the set then the corresponding product is taken to be zero.

Table IV. Time for the calculation of <u>one</u> phonon frequency as a function of the dimensions of the Hamiltonian matrix H and of the dielectric matrix ε.

dimension H	dimension ε	time (sec)
150	113	1376.1
150	137	1861.0
150	169	2637.7
150	181	3016.0
169	113	1655.1
169	137	2242.0
169	169	3176.1
169	181	3575.0
193	113	2032.7
193	137	2827.3
193	169	3909.9
193	181	4397.4

3. Find the address of each $\vec{G}'-\vec{G}$, put it equal to zero if the vector does not belong to the first 193. Care must be taken that the element of C_k^n with zero index contains the value zero.
4. Gather the elements $C_k^n(\vec{G}'-\vec{G})$ using an index list.
5. Calculate $T_{kk'}^{nn'}(\vec{G})$ with the special call Q8SDOT, controlled by a bit vector.

The run time of the complete phonon program is determined by the dimension of the Hamiltonian matrix H and by the dimension of the dielectric matrix ε. In table IV timing results for the complete phonon calculation are given for different dimensions of H and ε. All runs mentioned in table IV were made on a 1-pipe, 1 Mword CYBER 205 (located at the K.F.A. Karlsruhe, F.R.G.).

From table IV it can be derived that the run time depends on the dimension of the Hamiltonian matrix (N) as $N^{1.55}$ and on the dimension of the dielectric matrix (M) as $M^{1.63}$.

IV. DISCUSSION OF THE RESULTS

In the present stage of our work an important objective is the investigation of the convergence of the phonon frequencies as a function of
- number of plane waves used in the expansion Eq. (15)
- the dimension of the linear response matrices.
This investigation is very time-consuming and could not have been done without the use of a supercomputer.

1. Convergence as a Function of the Dimension of ε

Table V shows results for the phonon frequencies as a function of the dimension of ε.

It is seen that the convergence is relatively slow. Effects of higher reciprocal lattice vectors (beyond 181) can be taken into account by adding a diagonal block to the dielectric matrix [19]. Beyond the given dimension the elements of $\varepsilon(\vec{q},\vec{G},\vec{G}')$ are approximated by

$$\varepsilon(\vec{q},\vec{G},\vec{G}') = \delta_{GG'} \left[1 + \frac{16\pi \, \rho(o)}{|\vec{q}+\vec{G}|^4}\right] \tag{19}$$

with $\rho(o)$ the zeroth-order Fourier component of the charge density. The lower diagonal block does not mix with the upper diagonal block (the one that is calculated exactly) upon inverting the matrix, so that the

170

Table V. Phonon frequencies (in THz) for different
         dimensions of the dielectric matrix ε. Also
         shown are the experimental values.

dimension ε	Γ	TO(X)	LOA(X)	TA(X)
89	22.40	19.31	17.41	6.11
113	18.18	16.75	13.97	3.68
137	18.74	17.28	14.36	4.38
169	18.77	17.22	14.39	4.75
181	18.67	17.12	14.33	4.55
Experiment [18]	15.5	13.9	12.3	4.5

lattice sums in the dynamical matrix (see Eq. (8) and
(9)) can be summed to convergence.

2. Convergence as a Function of the Number of Plane Waves in the Basis

Table VI shows results for the phonon frequencies
as a function of the number of plane waves included in
the expansion Eq. (15). As the convergence is relatively slow an extrapolation was made where the phonon
frequency was supposed to be inversely proportional to
the number of plane waves in the basis, i.e.:

Table VI. Phonon frequencies of Si in terms of the
          dimension of the Hamiltonian matrix H, together
          with the extrapolated and experimental values
          (in THz).

dimension ε	Γ	TO(X)	LOA(X)	TA(X)
137	18.89	17.50	14.51	4.39
150	18.74	17.28	14.36	4.38
169	18.54	17.07	14.20	4.39
193	18.46	17.05	14.29	4.51
Extrapolation	17.10	14.93	12.81	4.16
Experiment [18]	15.5	13.9	12.3	4.5

$$\omega = A \frac{1}{N} + B \tag{20}$$

with ω the phonon frequency and N the number of plane waves. The constants A and B are determined using results of 137 and 150 plane waves. The resulting frequencies for an infinite number of waves are also shown in table VI, together with the experimental values.

3. The Dispersion Curves

The dispersion relations of Silicon in the Δ-, Λ- and Σ-directions are shown in Figure 2. These results are obtained from a calculation where both the diagonal block and the extrapolation is used.

A meaningful overall agreement with experiment is obtained. The discrepancy averaged over all branches and wave vectors is 14%. Generally, the acoustic

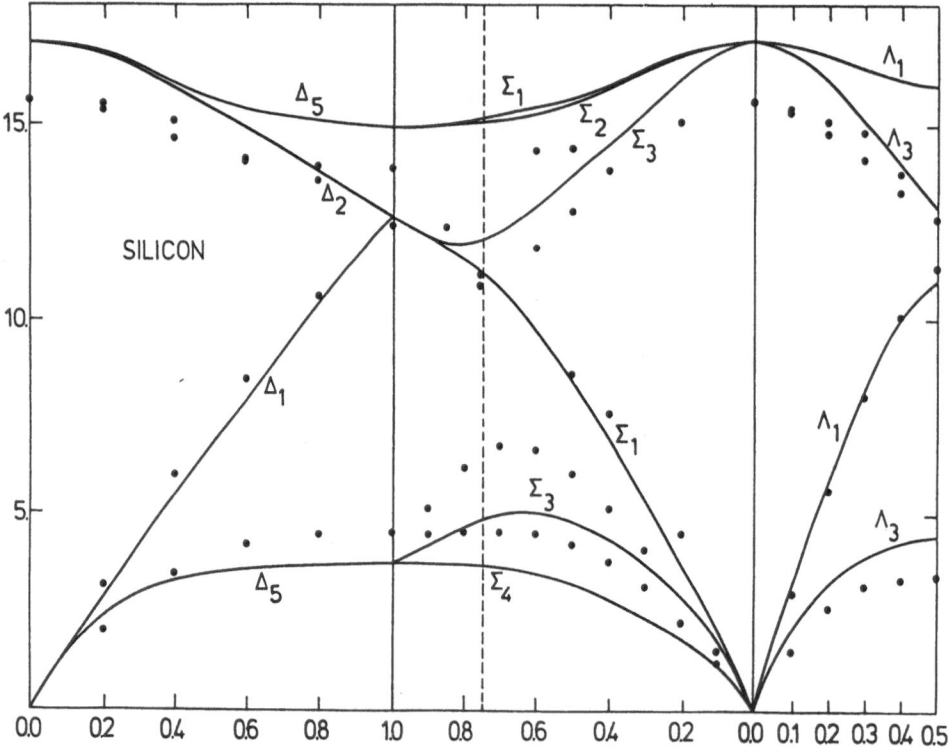

Fig. 2. Calculated phonon frequencies of Silicon in Δ-, Λ- and Σ-directions (in THz). The experimental data [18] are also shown.

branches are in better agreement with experiment than the
optical branches, while the optical modes are consistently too high. It must be emphasized that these curves
were obtained purely from first principles, i.e. no
adjustable parameters were used.

V. CONCLUSIONS

In this paper it is shown that for the ab-initio
calculation of the phonon dispersion curves of covalent
semiconductors in the dielectric screening formalism
the use of a vector computer is necessary.

As far as the physics is concerned the following
conclusions can be drawn:
1. There are three consistency requirements:
   - the electron-ion potentials in the Hamiltonian and
     in the dynamical matrix should be identical. There-
     fore, a self-consistent energy band calculation is
     necessary
   - the same exchange-correlation approximation should
     be made in the Hamiltonian and in the linear response
     theory
   - the total energy of the crystal should be minimal-
     ized with respect to the lattice spacing. As a
     consequence the crystal will be free of stress.
2. It is very important to use a sufficiently large
   basis. For the calculation of the polarizability
   matrix exactly, the same basis should be used. This
   means that all the calculated conduction bands should
   be taken into account. Even then an extrapolation to
   an infinite basis is necessary.

ACKNOWLEDGMENTS

The numerical work was performed on a CDC CYBER 205
with a grant from the "Supercomputer Project" by the
N.F.W.O. (National Fund for Scientific Research, Belgium).
The authors also would like to thank Dr. V.E. Van Doren
who is also involved in the research.

REFERENCES

[1] M.T. Yin and M.L. Cohen, Phys. Rev. Lett. 45:1004
    (1980); Phys. Rev. B25:4317 (1982).

R.M. Martin and K. Kunc, in: "Ab-Initio Calculation
of Phonon Spectra", J.T. Devreese, V.E. Van Doren,
P.E. Van Camp, eds., Plenum Press, New York (1983),
p. 49.

[2] R.M. Pick, M.H. Cohen and R.M. Martin, Phys. Rev.
B2:910 (1970).
P.E. Van Camp, V.E. Van Doren and J.T. Devreese, in:
"Ab-Initio Calculation of Phonon Spectra", J.T.
Devreese, V.E. Van Doren, P.E. Van Camp, eds.,
Plenum Press, New York (1983), p. 25.

[3] P.E. Van Camp, V.E. Van Doren and J.T. Devreese,
Phys. Rev. Lett. 42:1224 (1979); Phys. Stat. Sol.
(b) 93:483 (1979).

[4] W. Kohn and L.J. Sham, Phys. Rev. 140:A1133 (1965).

[5] P.E. Van Camp, V.E. Van Doren and J.T. Devreese,
Phys. Rev. B24:1096 (1981).

[6] J.C. Slater, Phys. Rev. 81:385 (1951).

[7] E.P. Wigner, Phys. Rev. 46:1002 (1934).

[8] D.M. Ceperley and B.J. Alder, Phys. Rev. Lett. 45:
566 (1980).

[9] F.A. Johnson, Proc. Roy. Soc. A310:101 (1969).

[10] P.E. Van Camp, V.E. Van Doren and J.T. Devreese,
Phys. Stat. Sol. (b) 110:K133 (1982).

[11] P.E. Van Camp, V.E. Van Doren and J.T. Devreese,
unpublished results.

[12] M. Schlüter, J.R. Chelikowsky, S. Louie and M.L.
Cohen, Phys. Rev. B12:4200 (1975).

[13] W. Topp and J. Hopfield, Phys. Rev. B7:1295 (1973).

[14] J. Donahue, "The Structure of the Elements", Wiley,
New York (1972).

[15] J. Ihm and M.L. Cohen, Phys. Rev. B21:1527 (1980).

[16] V. Strassen, Num. Math. 19:354 (1969).

[17] J.H. Wilkinson, "The Algebraic Eigenvalue Problem",
Oxford University Press, London (1965).

[18] G. Nilson and G. Nelin, Phys. Rev. B6:3777 (1972). 9

[19] P.E. Van Camp, V.E. Van Doren and J.T. Devreese, in:
"Proceedings of the 6th General Conference 'Trends
in Physics' of the European Physical Society,
Prague", J. Janta and J. Pantoflicek, eds. (1984),
p. 486.

NUMERICAL SOLUTION OF THE TIME-DEPENDENT HARTREE-FOCK

EQUATION IN THE ELECTRON GAS[x]

F. Brosens[o] and J.T. Devreese[oo]

Department of Physics, University of Antwerp
(UIA), Universiteitsplein 1, B-2610 Antwerpen-
Wilrijk, Belgium

I. INTRODUCTION

In this paper, we present the outline of a numerical
procedure which we are developing on a CYBER 205, and
which will allow to calculate the dielectric function of
the electron gas, including dynamical exchange effects.

In Section II, we first briefly discuss the homogeneous electron gas model, and its relevance for the study
of the electronic properties of solids.

In Section III, the derivation of the time-dependent
Hartree-Fock (TDHF) equation and the underlying basic
assumptions will be sketched. Some previous results
will be summarized, as obtained from a variational procedure [1] which we developed to handle this equation
approximately. The importance of the dynamical exchange
effects, as revealed by this variational method, gave the
basic motivation for solving the TDHF equation exactly.

Finally, in Section IV, a numerical solution procedure for the TDHF equation will be discussed and some

---

[x] Supported by the Supercomputer Project of the National Fund for Scientific Research, Belgium.
[o] Research Associate of the National Fund for Scientific Research, Belgium.
[oo] Also at: University of Antwerp (RUCA), Belgium and
University of Technology, Eindhoven, The Netherlands.

vectorization aspects for the CYBER 205 will be examined in more detail. Also some preliminary results will be presented.

II. THE HOMOGENEOUS ELECTRON GAS MODEL

The calculation of many properties of simple metals is based on the "jellium model". In this model, one merely studies the interaction between the electrons, whereas the lattice of positive metal ions is replaced by a rigid uniform background. The large amount of research on this hypothetical system is not only due to its conceptual simplicity. The main reason for the persistent interest lies in the fact that in many metals, the conduction electrons are almost homogeneously distributed in space. Furthermore, many theories for inhomogeneous systems use the jellium model as a reference system for further investigations.

The only equilibrium parameter of this model is the electron density n, which is often expressed in the Wigner-Seitz parameter r_s, which is the radius of a sphere, in units of the Bohr radius a_o, the volume of which is the mean volume available per electron:

$$\frac{1}{n} = \frac{4\pi}{3} r_s^3 a_o^3 \qquad\qquad (II.1)$$

A basic quantity for the study of the homogeneous electron gas, is the frequency- and wave vector-dependent longitudinal dielectric function $\varepsilon(q,\omega)$, which not only allows the study of the dielectric response, but also to calculate several other electron gas properties, e.g. the ground state energy, the dynamical structure factor $S(q,\omega)$, the pair correlation function $g(r)$, ...

Since the pioneering work [2] of Lindhard, Bohm, Pines, Nozières, ..., the dielectric function in the so-called Random Phase Approximation (RPA), has been the standard reference basis for further improvements. In this approximation, one calculates the response of the electrons, subjected to an external field, assuming that each electron moves in the Hartree field of all the other electrons.

Denoting the Fourier components of the external applied field by $\phi_{q,\omega}$, and the induced electron density by $n_{q,\omega}$, the dielectric function $\varepsilon(q,\omega)$ is defined as the ratio between the externally applied potential and the total potential, seen by a unit test charge:

$$e\phi_{q,\omega} + \frac{4\pi e^2}{q^2} n_{q,\omega} = \frac{e\phi_{q,\omega}}{\varepsilon(q,\omega)}$$ (II.2)

In the RPA, the dielectric function can be calculated analytically, but we only give an integral representation, which is useful for reference later on:

$$\varepsilon^{RPA}(q,\omega) = 1 + Q_o(q,\omega)$$ (II.3)

In this expression, the Lindhard function $Q_o(q,\omega)$ is given by

$$Q_o(q,\omega) = \frac{4\pi e^2}{q^2} \int d^3p \, \frac{N_{\vec{q}}(\vec{p})}{\omega^+ - \vec{p}\cdot\vec{q}/m}$$ (II.4)

where $\omega^+ = \omega+i\delta$ accounts for the adiabatic switching of external perturbations, and $N_{\vec{q}}(\vec{p})$ is a geometrical factor, related to the equilibrium distribution function of the electron gas:

$$N_{\vec{q}}(\vec{p}) = \frac{1}{\hbar} \, [f^\circ(\vec{p} + \frac{\hbar\vec{q}}{2}) - f^\circ(\vec{p} - \frac{\hbar\vec{q}}{2})]$$ (II.5)

where

$$f^\circ(p) = \frac{2}{(2\pi\hbar)^3} \qquad \text{if } p \leqslant p_F$$

$$= 0 \qquad \text{if } p > p_F$$ (II.6)

A large variety of approximations has been proposed to improve upon the RPA. In these approximations, one attempts to calculate the "local field correction" $G(q,\omega)$, which is usually introduced in the dielectric function in the following way:

$$\varepsilon(q,\omega) = 1 + \frac{Q_o(q,\omega)}{1-G(q,\omega)Q_o(q,\omega)}$$ (II.7)

The function $G(q,\omega)$ is intended to describe the exchange and correlation hole, due to the exchange and correlation interaction.

Among others, two important topics show the relevance of a detailed knowledge of the dielectric function (and therefore of $G(q,\omega)$), namely the energy loss of X-rays and fast electrons, when passing through a thin

177

metal slab, and the derivation of the effective potential in the ab-initio description of metals.

1. Energy Loss of X-Rays and Fast Electrons

The energy loss of X-rays and fast electrons, when scattered on a metal target, is one of the most direct sources of information on the dielectric function, since the differential cross section for scattering with given energy loss under a given solid angle, is proportional to the imaginary part of the inverse dielectric function $\varepsilon(q,\omega)$ (i.e. to the dynamical structure factor). The wave vector $\vec{q}$ and the frequency ω are the difference in wave vector and frequency between the incoming and the outgoing beam.

In a metal, two kinds of bulk excitations can be distinguished, namely the plasmons and the single-particle excitations. The plasmons are collective excitation modes of the electron gas, due to the long range Coulomb interaction. In the long-wavelength limit, their oscillation frequency is the classical plasma frequency ω_p:

$$\omega_p^2 = \frac{4\pi n e^2}{m} \qquad (II.8)$$

In the differential cross section as a function of frequency for fixed wave vector q, the inelastic scattering on the plasmons shows a peak at the plasmon frequency $\omega_p(q)$. Due to the damping by the single-particle excitations (i.e. the creation of electron-hole pairs), the plasmon peak is broadened in the appropriate energy range (i.e. in the particle-hole continuum).

In the long wavelength limit, the RPA succeeds basically in explaining the dispersion of the energy of the collective excitations, and simultaneously to combine the ideas of screening, collective excitations and single-particle excitations. However, with increasing wave vector, the RPA predictions for the plasmon dispersion become less satisfactory as compared to the experimental data [3].

This failure of the RPA in the short-range description of the electron-electron interactions is related to the neglect of exchange and correlation in the response. It is also reflected in the fact that the pair correlation function g(r), as obtained from the RPA dielectric function, becomes negative near the origin [4] for metallic densities.

2. The Effective Potential in Solids

The dielectric function of the homogeneous electron gas is also important for the construction of the effective potential in quantum-mechanical studies of solids, based on the Hohenberg-Kohn theorems [5]. Without entering into details, these theorems can be summarized as follows:
- The ground state energy of a (non degenerate) many-electron system is a unique functional of the density, which means that there exists a one-to-one correspondence between the electron density distribution and the externally applied potential (apart from an arbitrary constant);
- This energy functional of the density reaches its minimum at the correct density.

In general, the universal energy functional is unknown. But in the case of an electron system of almost constant density, Hohenberg and Kohn derived the coefficients of the expansion of the functional to second order in the density fluctuations. A basic ingredient in this expansion is the dielectric function of the homogeneous electron gas.

Based on the Hohenberg-Kohn theorem, Kohn and Sham [6] constructed a set of self-consistent single-particle Schrödinger equations, where the effective potential is essentially the functional derivative of the universal energy functional with respect to the electron density. Since the functional is unknown, as mentioned above, also the effective potential is not known exactly, and one has to rely on approximations (e.g. the local-density approximation (LDA), in which one assumes that the energy density functional at any point in space, equals the energy density of the homogeneous electron gas with the uniform density replaced by the local density at that point in space). However, in systems of almost constant density, this effective potential is known to first order in the density fluctuations in terms of the dielectric function of the homogeneous electron gas. Therefore, in such systems, the knowledge of the dielectric function would allow to give an explicit expression for the effective potential to first order in the density fluctuations.

III. THE TIME-DEPENDENT HARTREE-FOCK (TDHF) EQUATION

In the previous section, a local-field correction $G(q,\omega)$ was formally introduced in the dielectric function (see Eq. II.7) to account for exchange and correlation effects in the electronic response of the electron gas. But its exact evaluation would require the solution of the many-body problem. Even with the present day super-computers, this objective seems to be unrealistic.

In principle, dynamical exchange effects alone (neglecting Coulomb correlations between electrons of antiparallel spin) are described by the TDHF equation. But also for this equation, the mathematical complexity until recently hampered an exact solution. Only in the last year, we were able to develop a procedure which seems to be appropriate to find the solution numerically, as will be discussed in section V.

In the present section, we will not give the full derivation of the TDHF equation, but only discuss the basic assumptions. A derivation in the notations used here, and a more detailed analysis, can be found in the papers, mentioned under ref. [1].

For the derivation of the TDHF equation, one applies an external potential with Fourier components $e\phi_{q,\omega}$, giving rise to the Fourier components $n_{q,\omega}$ of the induced density in the electron gas. From the definition (Eq. II.2), the dielectric function can be derivated if the induced density is calculated.

This derivation can e.g. be expressed in terms of the Wigner distribution function $f_\sigma(\vec{p},\vec{q},\omega)$, where σ and $\vec{p}$ denote the spin and the momentum of the electrons under consideration, and $\vec{q}$ and ω are the wave vector and the frequency of the perturbation. The Wigner distribution function is the quantum analogue of the classical Boltzmann distribution function (see e.g. ref. [7]). For instance, the induced electron density is given by

$$n_{\vec{q},\omega} = \sum_\sigma \int d^3p \; f_\sigma(\vec{p},\vec{q},\omega) \qquad\qquad (III.1)$$

Deriving then the equations of motion for the Wigner distribution, and applying the Hartree-Fock de-coupling scheme in the electron dynamics (see ref. 1) one obtains the following integral equation:

$$f_\sigma(\vec{p},\vec{q},\omega) = \frac{-\frac{1}{2} N_{\vec{q}}(\vec{p}) U_{\vec{q},\omega} + X_\sigma(\vec{p},\vec{q},\omega)}{\omega^+ - \vec{p}\cdot\vec{q}/m} \qquad \text{(III.2a)}$$

$$U_{\vec{q},\omega} = e\phi_{\vec{q},\omega} + \frac{4\pi e^2}{q^2} \sum_\sigma \int d^3p\; f_\sigma(\vec{p},\vec{q},\omega) \qquad \text{(III.2b)}$$

$$X_\sigma(\vec{p},\vec{q},\omega) = \int d^3p' \frac{2\pi e^2 \hbar^2}{|\vec{p}-\vec{p}'|^2} [\, N_{\vec{q}}(\vec{p}) f_\sigma(\vec{p}',\vec{q},\omega) -$$

$$N_{\vec{q}}(\vec{p}') f_\sigma(\vec{p},\vec{q},\omega) \,] \qquad \text{(III.2c)}$$

In these equations, $U_{\vec{q}\omega}$ denotes the Hartree potential
from the induced electron density, and $X_\sigma(\vec{p},\vec{q},\omega)$ accounts
for the exchange effects in the dynamics. The geometric-
al factor $N_{\vec{q}}(\vec{p})$ also occurs in the expression for the
RPA dielectric function, and is given by equations (II.5)
and (II.6).

It should be emphasized that the RPA dielectric
function is readily derived from the TDHF equation
(III.2) if one neglects the exchange contribution
$X_\sigma(\vec{p},\vec{q},\omega)$. One then obtains the Lindhard distribution
function:

$$f^L_\sigma(\vec{p},\vec{q},\omega) = -\frac{1}{2} \frac{e\phi_{\vec{q},\omega}}{1+Q_0(q,\omega)} \frac{N_{\vec{q}}(\vec{p})}{\omega^+ - \vec{p}\cdot\vec{q}/m} \qquad \text{(III.3)}$$

Using the Lindhard distribution function to calculate
the induced density (Eq. III.1), and applying the
definition (II.2) of the dielectric function, one then
ends up with the RPA dielectric function.

IV. VARIATIONAL SOLUTION OF THE TDHF EQUATION

As mentioned in the previous section, the exchange
contribution $X_\sigma(\vec{p},\vec{q},\omega)$ to the TDHF equation leaves little
hope to solve the integral equation analytically. In
order to estimate the effects of the exchange interaction on the dielectric function, we developed a variational procedure, which consisted in the construction of a
functional $F[f_\sigma(\vec{p},\vec{q},\omega)]$ with the property that the
equation of motion (III.2) follows by imposing that
$F[f_\sigma(\vec{p},\vec{q},\omega)]$ is stationary with respect to variations
in the Wigner distribution function. This extremum
condition allows to construct variational approximations

with trial Wigner distribution functions. Mathematical
details can be found in ref. [1]. As a result, a
variational approximation for the dielectric function,
and thus for $G(q,\omega)$ was obtained:

$$G^{var}(q,\omega) = \frac{4\pi e^2}{q^2} \frac{2\pi e^2 \hbar^2}{Q_o(q,\omega)^2} \int d^3p \int d^3p' \frac{N_{\vec{q}}(\vec{p})N_{\vec{q}}(\vec{p}')}{|\vec{p}-\vec{p}'|^2}$$

$$\times \frac{1}{\omega^+ - \vec{p}\cdot\vec{q}/m} [\frac{1}{\omega^+ - \vec{p}'\cdot\vec{q}/m} - \frac{1}{\omega^+ - \vec{p}\cdot\vec{q}/m}] \qquad (IV.1)$$

To the best of our knowledge, this equation was the
first expression for $G(q,\omega)$ which was explicitly evaluated as a function of frequency and wave vector. For a
discussion of the numerical and analytical evaluation
techniques, as well as for an extensive discussion, we
refer to the original papers. In this paper, we only
summarize the main results, obtained from this variational local-field correction:
1) $G^{var}(q,\omega)$ turns out to be a universal function of
   q/k_F and $\hbar\omega/E_F$ for all densities. This "scaling" law

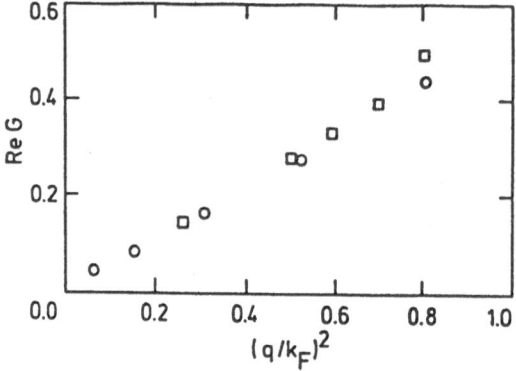

Fig. 1. Fitted values for the real part of $G(q,\omega)$ at the
        experimental plasmon frequencies (ref. [8]) for
        Al (squares) and Na (circles), giving evidence
        for the "scaling" property of $G(q,\omega)$ with respect
        to density, as suggested by the variational
        solution of the TDHF equation.

seems to be confirmed by experimental data from the plasmon spectrum in Na and Al. This is shown in Fig. 1, where the real part of $G(q,\omega)$ at the plasmon spectrum for several wave vectors is plotted, as obtained by fitting the maximum of the structure factor to the plasmon energy [8]. Despite the fact that the valence electron densities in Na and Al differ by a factor of 8, the fitted values for $G(q,\omega)$ at the plasmon frequency remarkably coincide for the different values of q/k_F.

2) The frequency dependence in $G^{var}(q,\omega)$ is rather pronounced. In order to satisfy several sum rules in the electron gas, this frequency dependence is absolutely required.

3) Dynamical exchange effects substantially lower the calculated plasmon frequency at finite wave vector as compared to the RPA, as is illustrated in several of the papers mentioned in ref. [1]. This tendency appreciably improves the agreement with experimental data.

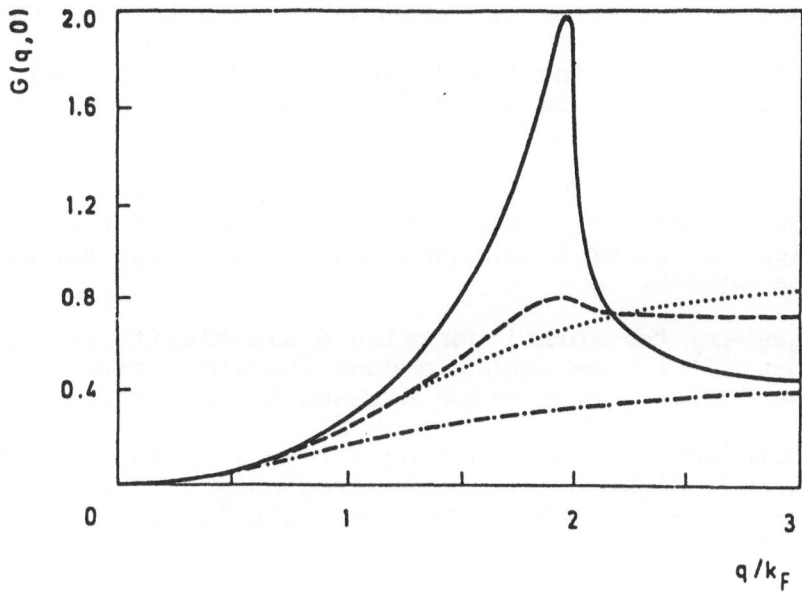

Fig. 2. Local field correction $G(q,o)$ in the static limit as obtained by the variational solution of the TDHF equation (full line), compared to the approximations of Hubbard [10] (dash-dotted line), Toigo and Woodruff [11] (dashed line) and Vashishta and Singwi [12] (dotted line) at $r_s = 3$.

4) In the static limit, the variational local-field correction exhibits a rather pronounced peak near $q = 2k_F$, as shown in Fig. 2. From spin- and charge-density wave considerations, Overhauser [9] argued that in this region a divergence should occur, but in most calculations of $G(q,o)$, this peak is weaker, or not present at all.

V. NUMERICAL SOLUTION OF THE TDHF EQUATION

In the previous sections, we presented some arguments, stressing the importance of dynamical exchange effects in the dielectric response of the electron gas, and the need for an exact solution of the TDHF equation. Indeed, without an exact solution, it is difficult to point out which effects are explicitly due to the exchange interaction, or artefacts of the approximations used. However, the solution of the TDHF equation is far from trivial, and requires the judicious use of a vector computer, since with a scalar computer one can hardly hope to obtain the solution for a sufficiently large set of parameter values within reasonable computation time.

Before discussing the numerical solution procedure, let us first summarize the difficulties, presented by the integral equation (III.2).

First of all, the TDHF equation is an integral equation in three variables (the components of the vector $\vec{p}$), which is rather time consuming, even if the kernel would be smooth.

However, the kernel contains a singularity of the form $|\vec{p}-\vec{p}'|^{-2}$ in the exchange contribution, resulting from the Coulomb interaction between the electrons.

Furthermore, a second singularity from the factor $(\omega^+-\vec{p}.\vec{q}/m)^{-1}$ might occur if the frequency of the external excitation lies in the range of the single-particle excitations.

Finally, this rather singular three-dimensional integral equation has to be solved several ten thousand times. Indeed, it depends on three parameters: the wave vector $\vec{q}$ and the frequency ω of the external excitation, and the density of the electron gas. (The density enters via the equilibrium distribution function, which is determined by the Fermi momentum.)

For a numerical treatment, one of course prefers to work with dimensionless variables. Since the Fermi units are quite natural in the electron gas, we introduce the new variables

$$\vec{k} = \vec{q}/k_F \; ; \; \vec{\xi} = p/\hbar k_F \; ; \; \nu = \hbar\omega/2E_F \tag{V.1}$$

Collecting then the remaining density dependent terms in a single parameter:

$$C = 2\pi^2 \, (\frac{9\pi}{4})^{1/3} \frac{1}{r_s} \tag{V.2}$$

a much more elegant expression is obtained instead of (III.2).

However, since $f_\sigma(\vec{p},\vec{q},\omega)$ is a singular function because of the denominator $(\omega-\vec{p}.\vec{q}/m)$, it is more appropriate for numerical purposes to consider a physically less appealing, but smooth function. After some algebra, one realizes that a numerically much simpler integral equation is obtained by introducing the new unknown function

$$b(\vec{\xi},\vec{k},\nu) = \frac{1}{e\phi_{\vec{q},\omega}} \, C \, 4\pi e^2 \hbar^3 k_F \, \varepsilon(k,\nu) \, (\nu^+ - \Lambda_k(\xi))$$

$$f_\sigma(\vec{p} + \frac{\hbar}{2}\,\vec{q},\vec{q},\omega)$$

where $\Lambda_{\vec{k}}(\vec{\xi})$ results from the Hartree-Fock one-electron energies, and is given by

$$\Lambda_{\vec{k}}(\vec{\xi}) = \frac{k^2}{2} + \vec{\xi}.\vec{k} - \frac{2\pi}{C} \, [T(\vec{\xi}+\vec{k}) - T(\vec{\xi})] \tag{V.3}$$

with

$$T(\vec{\xi}) = \frac{1-\xi^2}{2\xi} \, \ln\left|\frac{1+\xi}{1-\xi}\right| \tag{V.4}$$

Using these algebraic transformations, the integral equation (III.2) takes the form:

$$b(\vec{\xi},\vec{k},\nu) = 1 - \frac{1}{C} \int_V d^3\xi' \, \{\frac{1}{|\vec{\xi}+\vec{\xi}'+\vec{k}|^2} \frac{b^*(\vec{\xi}',\vec{k},-\nu)}{[(-\nu)^+ - \Lambda_{\vec{k}}(\vec{\xi}')]^*}$$

$$+ \frac{1}{|\vec{\xi}-\vec{\xi}'|^2} \frac{b(\vec{\xi}',\vec{k},\nu)}{\nu^+ - \Lambda_{\vec{k}}(\vec{\xi}')}\} \; \text{for} \; \vec{\xi} \in V \tag{V.5}$$

where the integration volume and the definition region

are defined by the condition

$$\vec{\xi} \in V \iff |\vec{\xi}| \leqslant 1 \quad \text{and} \quad |\vec{\xi}+\vec{k}| \geqslant 1 \tag{V.6}$$

The dielectric function can be directly expressed in terms of this unknown function $b(\vec{\xi},\vec{k},\nu)$:

$$\varepsilon(k,\nu) = 1 - \frac{1}{k^2} \frac{2}{c} \int_V d^3\xi \left\{ \frac{b^\star(\vec{\xi},\vec{k},-\nu)}{[(-\nu)^+ - \Lambda_{\vec{k}}(\vec{\xi})]^\star} + \frac{b(\vec{\xi},\vec{k},\nu)}{\nu^+ - \Lambda_{\vec{k}}(\vec{\xi})} \right\} \tag{V.7}$$

Because these equations obviously show cylindrical symmetry with respect to rotations around the excitation wave vector $\vec{k}$, the azimuthal angle can be integrated out, and one is left with an integral equation for a two-dimensional vector:

$$b(\vec{\xi},\vec{k},\nu) = 1 - \frac{2\pi}{c} \int_S d^2\xi \left\{ \frac{\xi'_\perp \, j(\vec{\xi},-\vec{\xi}'-\vec{k})}{[(-\nu)^+ - \Lambda_{\vec{k}}(\vec{\xi}')]^\star} b^\star(\vec{\xi}',\vec{k},-\nu) \right.$$

$$\left. + \frac{\xi'_\perp \, j(\vec{\xi},\vec{\xi}')}{\nu^+ - \Lambda_{\vec{k}}(\vec{\xi}')} b(\vec{\xi}',\vec{k},\nu) \right\} \text{for } \vec{\xi} \in S \tag{V.8}$$

where

$$\vec{\xi} \in S \iff |\vec{\xi}| \leqslant 1 \quad \text{and} \quad |\vec{\xi}+\vec{k}| \geqslant 1 \tag{V.9}$$

$\vec{\xi}'_\perp$ is the component of $\vec{\xi}$, orthogonal to $\vec{k}$

and

$$j(\vec{\xi},\vec{\xi}') = \left[((\xi_z - \xi'_z)^2 + \xi_\perp^2 + \xi_\perp'^2)^2 - 4\xi_\perp^2 \xi_\perp'^2 \right]^{-1/2} \tag{V.10}$$

For the dielectric function one then obtains:

$$\varepsilon(k,\nu) = 1 - \frac{2}{k^2} \frac{2\pi}{c} \int_S d^2\xi \left\{ \frac{\xi'_\perp \, b^\star(\vec{\xi},\vec{k},-\nu)}{[(-\nu)^+ - \Lambda_{\vec{k}}(\vec{\xi})]^\star} \right.$$

$$\left. + \frac{\xi'_\perp \, b(\vec{\xi},\vec{k},\nu)}{\nu^+ - \Lambda_{\vec{k}}(\vec{\xi})} \right\} \tag{V.11}$$

In cartesian coordinates $(\xi_z,\xi_\perp)$ with the z-axis along k, the surface S, defined in (V.9), becomes a semicircle of unit radius in the half plane $\xi_\perp > 0$, with the center in the origin $\xi_z = \xi_\perp = 0$, from which in the case $k \leqslant 2$ an overlap region has to be substracted with a similar semicircle centered around $\xi_z = -k$, $\xi_\perp = 0$.

1. Construction of a Set of Linear Equations

Since the two-dimensional integral equation, derived above, has to be solved for a large number of values of the parameters k, ν and c, efficient procedures have to be developed, which are easily vectorizable.

It can be shown analytically that the function $b(\vec{\xi}, \vec{k}, \nu)$ is smooth, as compared to the kernel. As a consequence, sufficiently small subdomains S_j of the integration region S can be defined, in which the unknown function can be replaced by the mean value in these subdomains. Defining thus:

$$b_J(k,\nu) = \frac{1}{S_J} \int_{S_J} d^2\xi \; b(\vec{\xi}, \vec{k}, \nu) \qquad (V.12)$$

these mean values can be taken out of the integrations, and one obtains a set of linear equations in the unknowns $b_J(k,\nu)$ instead of an integral equation:

$$b_I(k,\nu) = 1 - \frac{2\pi}{c} \sum_J [F_{I,J}(k,\nu) b_J(k,\nu) + G^*_{I,J}(k,-\nu)$$

$$b^*_J(k,-\nu)] \qquad (V.13)$$

with

$$F_{I,J}(k,\nu) = \int_{S_J} d^2\xi \; \frac{\xi_\perp \; g_I(\vec{\xi})}{\nu^+ - \Lambda_{\vec{k}}(\vec{\xi})} \qquad (V.14)$$

$$G_{I,J}(k,\nu) = \int_{S_J} d^2\xi \; \frac{\xi_\perp \; g_I(-\vec{\xi}-\vec{k})}{\nu^+ - \Lambda_{\vec{k}}(\vec{\xi})} \qquad (V.15)$$

where

$$g_I(\vec{v}) = \frac{1}{S_I} \int_{S_I} d^2\xi \; j(\vec{\xi}, \vec{v}) \qquad (V.16)$$

Along the same lines, the dielectric function becomes:

$$\varepsilon(k,\nu) = 1 - \frac{2}{k^2} \frac{2\pi}{c} \sum_J [H_J(k,\nu) b_J(k,\nu) + H^*_J(k,-\nu) b^*_J(k,-\nu)]$$

$$(V.17)$$

187

with

$$H_J(k,\nu) = \int_{S_J} d^2\xi \; \frac{\xi_1}{\nu^+ - \Lambda_{\vec{k}}(\vec{\xi})} \qquad (V.18)$$

In practice, the subdomains S_J were defined with the following scheme. The radius of the total integration domain was subdivided in equal parts, and the shells obtained were subdivided with respect to the polar angle in domains of equal surface. Each subdomain is then characterized by four data: ξ_J^{min}, ξ_J^{max}, θ_J^{min}, θ_J^{max}, as indicated in Fig. 3. These values are contained in a table. This shape of the subdomains has the advantage that they can easily be redefined in smaller subdomains of the same shape, depending on the smoothness of the solution. Such a refinement of the mesh only requires the modification of some values in the table, and the introduction of some new values. (Note that the ordering of the indices is irrelevant).

It should be emphasized that the case $k < 2$ introduces some complications (see V.9), but we do not go further into details. This case requires some extra careful bookkeeping, but does not alter the concept of the program.

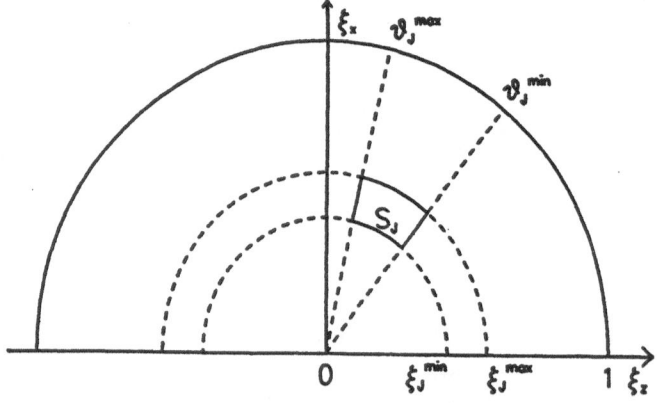

Fig. 3. Graphical illustration of the parametrizations of the subdomains S_J, as explained in the text.

The main problem left is thus the evaluation of $F_{I,J}(k,\nu)$, $G_{I,J}(k,\nu)$ and $H_J(k,\nu)$ in the equations (V.14-15,18). However, the evaluation of $F_{I,J}(k,\nu)$ and $G_{I,J}(k,\nu)$ requires the computation of the supplementary integral $g_I(\vec{v})$, with the integrand given in (V.10).

It should be noted that $g_I(\vec{v})$ can analytically be evaluated in terms of elliptic integrals. But it turns out that a faster procedure, which is easier to vectorize, is obtained by expressing the integral (V.16) in polar coordinates, performing the angular integration analytically, which gives inverse hyperbolic functions, and leaving the radial integration for numerical computation. The algebra involved is quite tedious but straightforward, and we do not discuss it in further detail.

To calculate the functions $F_{I,J}(k,\nu)$, $G_{I,J}(k,\nu)$ and $H_J(k,\nu)$, it should be noted that vectorization with respect to the subdomain subscripts I and J is quite easy to realize. Indeed, these indices only refer to the integration domain considered, the limits of which are contained in a table, as discussed above.

In order to vectorize also with respect to the parameter values k and ν, one has to develop the appropriate integration procedures, which we will discuss below. However, because in practice the number of subdomains is of order 100 to 500 and some 50 till 100 frequency values are to be considered, there is no use in vectoring also with respect to the wave vector k, since the maximum vector length on the CYBER 205 is of order 65000.

For the vectorization of $F_{I,J}(k,\nu)$, $G_{I,J}(k,\nu)$ and $H_J(k,\nu)$ with respect to the frequency ν, one should note that these integrals are of the form

$$C_J^{[f]}(k,\nu) = \int_{S_J} d^2\xi \, \frac{f(\xi)}{\nu^+ - \Lambda_{\vec{k}}(\vec{\xi})} \quad \text{with } f(\vec{\xi}) \text{ real} \quad (V.19a)$$

and where $f(\vec{\xi})$ shows no singularities in the integration domain. Therefore, one can proceed in two steps, to be vectorized separately:
1) Evaluate the imaginary part:

$$A_J^{[f]}(k,\nu) = \text{Im} \int_{S_J} d^2\xi \, \frac{f(\vec{\xi})}{\nu^+ - \Lambda_{\vec{k}}(\vec{\xi})} \quad (V.19b)$$

2) Evaluate the real part:

$$B_J^{[f]}(k,\nu) = \text{Re} \int_{S_J} d^2\xi \; \frac{f(\vec{\xi})}{\nu^+ - \Lambda_{\vec{k}}(\vec{\xi})} \qquad \text{(V.19c)}$$

using the Kramers-Kronig relation.

2. Evaluation of $A_J^{[f]}(k,\nu) = \text{Im} \int_{S_J} d^2\xi \; \dfrac{f(\vec{\xi})}{\nu^+ - \Lambda_{\vec{k}}(\vec{\xi})}$

For the computation of this integral, which is needed for $f(\vec{\xi})$ real and without singularities in the integration domain, one integration can be eliminated by using:

$$\text{Im} \lim_{\varepsilon \to 0^+} \int_{x_2}^{x_1} dx \; \frac{g(x)}{x - a - i\varepsilon} = \pi \, \delta(a) \quad \text{for } x_1 < a < x_2$$

It can be shown that $\Lambda_{\vec{k}}(\vec{\xi})$ (see eq. IV.3) is a monotonically decreasing function of the polar angle. Expressing the integrals in $A_J^{[f]}(k,\nu)$ in polar coordinates, the angular integration can thus be eliminated, giving:

$$A_J^{[f]}(k,\nu) = \pi \int_{\xi_J^{\min}}^{\xi_J^{\max}} \xi \, d\xi \int_{\theta_J^{\min}}^{\theta_J^{\max}} d\theta \; \Theta(\Lambda_{\vec{k}}(\xi, \theta_J^{\min}) - \nu)$$

$$\Theta(\nu - \Lambda_{\vec{k}}(\xi, \theta_J^{\max})) \, \delta(\nu - \Lambda_{\vec{k}}(\xi, \theta)) \, f(\xi, \theta) \quad \text{(V.20)}$$

where $\Theta(x) = 1$ for $x > 1$, and zero otherwise. The Θ-functions determine the range of frequency values ν for which $A_J^{[f]}(k,\nu)$ differs from zero.

As a first step, one thus determines numerically the minimum and maximum frequency values in each subdomain S_J:

$$\nu_J^{\max} = \text{Max}[\Lambda_{\vec{k}}(\vec{\xi})] \quad \text{for } \vec{\xi} \in S_J$$

$$\nu_J^{\min} = \text{Min}[\Lambda_{\vec{k}}(\vec{\xi})] \quad \text{for } \vec{\xi} \in S_J \qquad \text{(V.21)}$$

The integral (V.20) then becomes:

$$A_J^{[f]}(k,\nu) = \pi \int_{\xi_J^{min}}^{\xi_J^{max}} \xi d\xi \; \frac{f(\xi,\theta)}{\left| \frac{d}{d\theta} \Lambda_{\vec{k}}(\xi,\theta) \right|}_{\theta=\theta_o(k,\nu,\xi)}$$

$$\text{for } \nu_J^{min} < \nu < \nu_J^{max}$$

$$= 0 \qquad\qquad \text{otherwise} \qquad\qquad (V.22)$$

where $\theta_o(k,\nu,\xi)$ is the polar angle determined from the condition

$$\Lambda_{\vec{k}}(\xi,\theta_o(k,\nu,\xi)) = \nu \quad \text{for } \nu_J^{min} < \nu < \nu_J^{max} \qquad (V.23)$$

Since $\Lambda_{\vec{k}}(\vec{\xi})$ is a monotonically decreasing function with increasing polar angle, the numerical solution of (V.23) does not present any difficulty worth mentioning, as soon as one has decided at which values of ν and ξ the solution is to be found.

For easy use later in the program, we defined a sufficiently dense mesh of N_ν frequencies ν_{J,i_ν} in each subdomain S_J, with N_ν of order 10 or 20, and equidistantly distributed between ν_J^{min} and ν_J^{max}:

$$\nu_{J,i_\nu} = \nu_J^{min} + \frac{i_\nu-1}{N_\nu-1} (\nu_J^{max}-\nu_J^{min}) \text{ for } i_\nu = 1,2,\ldots N_\nu$$

$$(V.24)$$

and store these frequencies in a table.

Rember that for each of these frequencies, the integral (V.22) has to be computed, taking into account the condition (V.23), which determines minimum and maximum values of ξ for each frequency ν in each sub-domain S_J. This yields a table of values

ξ_{J,i_ν}^{min} = minimum value $\left.\vphantom{\begin{array}{c}a\\b\\c\end{array}}\right]$ of ξ for which (V.23) can be satisfied

ξ_{J,i_ν}^{max} = maximum value $\left.\vphantom{\begin{array}{c}a\\b\\c\end{array}}\right\}$ in S_j for ν_{J,i_ν}

Because the differentiation of $\Lambda_{\vec{k}}(\xi,\theta)$ with respect to θ, as it appears in the denominator of (V.22) can be performed analytically, it can be verified that the Chebychev quadrature of the first kind is quite appropriate. Performing this integration with N_ξ points (with N_ξ of order 10), we thus define a mesh:

$$\xi_{J,i_\nu,i_\xi} = \frac{\xi^{max}_{J,i_\nu} + \xi^{min}_{J,i_\nu}}{2} + \frac{\theta^{max}_{J,i_\nu} - \theta^{min}_{J,i_\nu}}{2} \cos \frac{\pi(2i_\xi - 1)}{2N_\xi}$$

$$\text{for } i = 1,2,..N_\xi \tag{V.25a}$$

whereas in these points the condition (V.23) defines corresponding polar angles

$$A_k(\xi_{J,i_\nu,i_\xi}, \theta_{J,i_\nu,i_\xi}) = \nu_{J,i_\nu} \tag{V.25b}$$

Using the standard weighting factors for the Chebychev quadrature of the first kind, and incorporating the contribution of the denominator of (V.21) in these weighting factors:

$$w_{J,i_\nu,i_\xi}(k) = -\frac{\pi}{N_\xi} \frac{\pi}{k} \frac{1}{|\sin\theta_{J,i_\nu,i_\xi}|} \frac{1}{\left|1 - \frac{2}{Cv} \frac{dT(v)}{dv}\right|}_{v = \vec{k} + \vec{\xi}_{J,i_\nu,i_\xi}} \tag{V.25c}$$

the computation of $A_J^{[f]}(k,\nu)$ in the N_ν frequencies ν_{J,i_ν} is given by

$$A_{J,i_\nu}^{[f]}(k) \simeq \sum_{i_\xi=1}^{N_\xi} w_{J,i_\nu,i_\xi} f(\vec{\xi}_{J,i_\nu,i_\xi}) \tag{V.25d}$$

As already suggested by the notations used, the vectorization of this procedure is rather easy. It requires the storage of the appropriate values ξ_{J,i_ν,i_ξ}, θ_{J,i_ν,i_ξ} and w_{J,i_ν,i_ξ} in (rather lengthy) tables, and the computation of the function $f(\vec{\xi})$ in all the values (θ,ξ) in the tables. The result is a vector of values $A_{J,i_\nu}^{[f]}$ for each function $f(\vec{\xi})$ to be considered, which are given in Eq. (V.14-15) and (V.18).

Even with only 10 subdomains, and 10 frequencies per subdomain, the vector length of the imaginary part of the matrices $F_{I,J}(k,\nu)$ and $G_{I,J}(k,\nu)$ is already 25000. Similarly, with only 50 subdomains, 10 frequencies per subdomain and 10 mesh points for the Chebychev integration, the vectors ξ_{J,i_ν,i_ξ}, J,i_ν,i_ξ and w_{J,i_ν,i_ξ} have a length of 5000. Thus even with a very limited number of evaluation points for the integral equation,

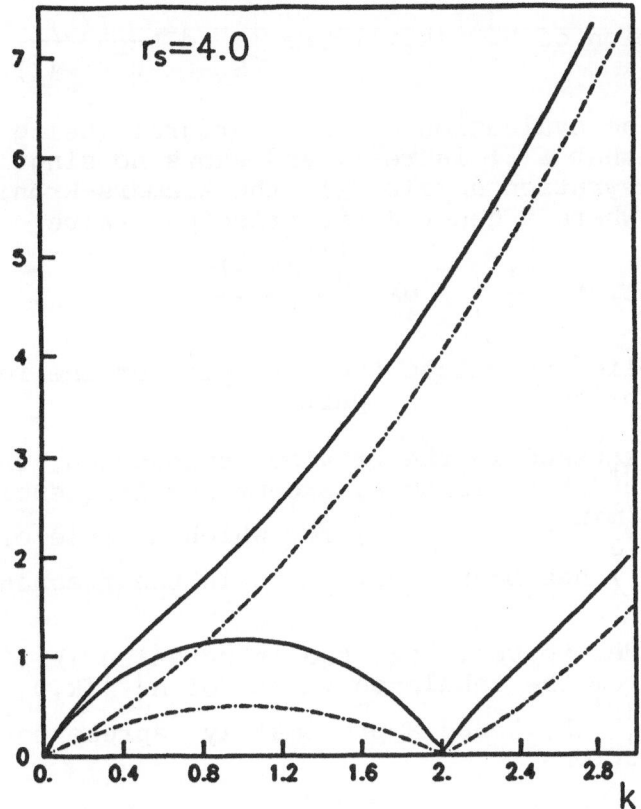

Fig. 4. The single-particle excitation spectrum boundaries as obtained in the RPA (dash-dotted line)
and including exchange corrections (full line).

a high degree of vectorization is realized by this
procedure.

It should be emphasized that this procedure at once
determines the frequencies for which the integral equation
(V.5), and therefore the dielectric function (V.7), has
an imaginary part different from zero, which corresponds
to determining the single-particle excitation spectrum
of the electron gas in the Hartree-Fock approximation.
As compared to the free-particle excitation spectrum
(used in our variational approach), the particle-hole
continuum in the TDHF approximation shows an increasing
deviation with increasing r_s (i.e. with decreasing density). The continuum obtained is shown in Fig. 4 for
$r_s = 4$.

3. Evaluation of $B_J^{[f]}(k,\nu) = \text{Re} \int_{S_J} d^2\xi \frac{f(\vec{\xi})}{\nu^+ - \Lambda_{\vec{k}}(\vec{\xi})}$

For the evaluation of this integral (using the condition that $f(\vec{\xi})$ is real, and shows no singularities in the integration domain S_J), the Kramers-Kronig relation (where f denotes the principal value)

$$B_J^{[f]}(k,\nu) = \frac{1}{\pi} \int_{-\infty}^{\infty} d\lambda \frac{A_J^{[f]}(k,\lambda)}{\lambda-\nu} \qquad (V.26)$$

can be applied to relate the real part of the required integral to its imaginary part.

As discussed in the previous subsection, the imaginary part $A_J^{[f]}(k,\nu)$ is zero, except for frequencies within the range $\nu_J^{min} < \nu < \nu_J^{max}$, for which a table of values $A_{J,i_\nu}^{[f]}(k,\nu_{i_\nu})$ has been constructed in the frequencies ν_{i_ν}.

In order to vectorize the integration (V.26), starting from the tabulated values of $A_J^{[f]}(k,\nu_{J,i_\nu})$, the function $A_J^{[f]}(k,\nu)$ was approximately represented by spline functions

$$\nu_{J,i_\nu} < \lambda < \nu_{J,i_\nu+1} : A_J^f(k,\lambda) \simeq a_{J,i_\nu} + b_{J,i_\nu}(\lambda-\nu_{J,i_\nu})$$

$$+ c_{J,i_\nu}(\lambda-\nu_{J,i_\nu})^2 + d_{J,i_\nu}(\lambda-\nu_{J,i_\nu})^3$$

$$(V.27)$$

where the expansion coefficients were determined by the constraints that the function itself, and its first and second derivative are continuous at the matching points of the interval.

It is well known that these constraints give rise to a tridiagonal set of linear equations, the vectorization of which has become standard. In the case under consideration, we could profit of the fact that this tridiagonal set of equations has to be solved for each subdomain S_J, allowing vectorization throughout the standard solution scheme for tridiagonal systems, by using descriptions instead of matrix elements.

Having determined the real part and the imaginary part of the integrals (V.14-15) and (V.18), the matrix elements of the set of linear equations (V.18), and the arrays to calculate the dielectric function (V.17) are completely determined. The set of linear equations was solved, using the LINPACK library package. The vectorization offered in LINPACK is not too satisfactory, since the CPU time in this particular case turns out to be comparable to the (completely scalar) subroutine CMLIN in the CERN library package. However, the solution of the set of equations only gives a minor contribution to the total CPU time, which is for more than 80% consumed by the evaluation of the matrix elements (even in the strongly vectorized version, as presented here).

VI. CONCLUDING REMARKS

The program, as sketched in the preceding sections, was first developed and tested on a VAX 11/780 in scalar version. Taking 49 subdomains, with 7 equidistant values of $\Lambda_{\vec{k}}$ per subdomain and 7 points in the Chebychev integration of the first kind, the calculation of the dielectric function in a single wave vector, frequency and density required about 210 seconds of CPU time. For the same program on the CYBER 205 in Karlsruhe, with automatic vectorization and with the unsafe optimizer, the computation time was reduced by a factor of 30. An extra gain factor of 28 was obtained by vectorizing the program, along the lines described above.

In Fig. 5, we present some results for the static value of the local-field correction $G(q,o)$ for various electron densities, and compare them to our previous variational results. For all values of r_s examined up to now, a remarkable agreement shows up between the numerical and the variational solution, except near $q = 2k_F$, where the numerically obtained $G(q,o)$ has a more pronounced peak. However, at present we cannot state whether there is a singularity present, or whether numerical inaccuracy is at the basis of this phenomenon. This question is under current investigation.

As far as the dynamical results are concerned, only a limited number of results has been obtained up to now. One of the major results is already presented in Fig. 4, indicating that the dynamical exchange effects considerably influence the single-particle excitation spectrum. This phenomenon was not accurately taken into account in our variational approach. This is presumably the reason why the $G^{var}(q,\omega)$ shows logarithmic singularities near the free-particle excitation boundaries, due to compressing the difference between the Hartree-Fock continuum into the free-particle continuum.

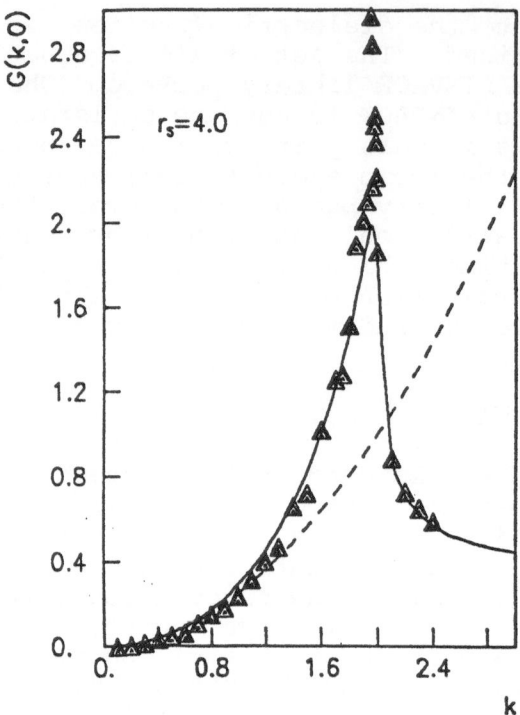

Fig. 5. Static values of the local field correction for
r_s = 4 as obtained from the numerical solution
of the TDHF equation (triangles) and compared to
the variational solution (full line). The
dashed line is the local field correction which
is assumed in the local density approximation.

However, in our present treatment, numerical accuracy problems seem to occur in the region where absorption
and emission of single-particle excitations almost balance
each other. This problem, as well as the possible
singularity of the static limit near q = $2k_F$, requires
analytical methods, which are in progress.

REFERENCES

[1] F. Brosens, L.F. Lemmens and J.T. Devreese, Phys.
Stat. Sol. (b) 74:45 (1976);
F. Brosens, J.T. Devreese and L.F. Lemmens, Phys.
Stat. Sol. (b) 81:99 (1977);

J.T. Devreese, F. Brosens and L.F. Lemmens, <u>Phys. Rev.</u> B21:1349 (1980);
F. Brosens, J.T. Devreese and L.F. Lemmens, <u>Phys. Rev.</u> B21:1363 (1980);
F. Brosens and J.T. Devreese, <u>in</u>: "Proceedings of the 1981 Advanced Study Institute on Electron Correlations in Solids, Molecules and Atoms", J.T. Devreese and F. Brosens, eds., Plenum Press, New York (1983), p. 143;
F. Brosens and J.T. Devreese, <u>Helv. Phys. Acta</u> 56: 223 (1983);
H. Nachtegaele, F. Brosens and J.T. Devreese, <u>Phys. Rev.</u> B28:6064 (1983);
F. Brosens and J.T. Devreese, <u>Phys. Rev.</u> B29:543 (1984).

[2] J. Lindhard, <u>Kong. Danske Vid. Selskab., Mat.-Fyss. Medd.</u> 24:No 8 (1954).
D. Bohm and D. Pines, <u>Phys. Rev.</u> 92:609 (1953).
D. Pines and P. Nozières, <u>"The Theory of Quantum Liquids"</u>, Vol. 1, Benjamin, New York (1966).

[3] D.M. Miliotis, <u>Phys. Rev.</u> B3:701 (1971).
P. Zacharias, <u>J. Phys.</u> F5:645 (1975).
H.J. Höhberger, A. Otto and E. Petri, <u>Solid State Commun.</u> 16:175 (1975).
P.M. Platzman and P. Eisenberger, <u>Phys. Rev. Lett.</u> 33:152 (1974).
P.E. Batson, C.H. Chen and J. Silcox, <u>Phys. Rev. Lett.</u> 37:937 (1976).
P.C. Gibbons, S.E. Schnatterly, J.J. Ritsko and J.R. Fields, <u>Phys. Rev.</u> B13:2451 (1976).
S.E. Schnatterly, <u>in</u>: "Solid State Physics 34", H. Ehrenreich, F. Seitz and D. Turnbull, eds., Academic Press, New York (1979).
S.E. Schnatterly, <u>in</u>: "Proceedings of the 1981 Advanced Study Institute on Electron Correlations in Solids, Molecules and Atoms", J.T. Devreese and F. Brosens, eds., Plenum Press, New York (1983).

[4] A.J. Glick and R.A. Ferrell, <u>Ann. of Phys.</u> 11:359 (1960).

[5] P. Hohenberg and W. Kohn, <u>Phys. Rev.</u> 136:B864 (1964).

[6] W. Kohn and L.J. Sham, <u>Phys. Rev.</u> 140:A1133 (1965).

[7] W.E. Brittin and W.E. Chappell, <u>Rev. Mod. Phys.</u> 34: 620 (1962).

[8] S.E. Schnatterly, <u>in</u>: "Proceedings of the 1981 Advanced Study Institute on Electron Correlations in Solids, Molecules and Atoms", J.T. Devreese and F. Brosens, eds., Plenum Press, New York (1983).

[9] A.W. Overhauser, <u>Phys. Rev.</u> B10:4918 (1974).

[10] J. Hubbard, <u>Proc. Roy. Soc.</u> A243:336 (1975).

[11] F. Toigo and T.O. Woodruff, <u>Phys. Rev.</u> B2:3958 (1970); 4:371 (1971).
[12] P. Vashishta and K.S. Singwi, <u>Phys. Rev.</u> B6:875 (1972).

AN OVERVIEW OF COMPUTING AT LOS ALAMOS°

Robert H. Ewald

Division Leader
Computing and Communications Division
Los Alamos National Laboratory
Los Alamos, New Mexico, U.S.A.

ABSTRACT

The Los Alamos National Laboratory operates one of
the largest scientific and engineering computing facilities in the world to support its national defence and
other programs. This facility, as of the summer of 1984,
had six Cray-computers, five CDC-7600-class machines, and
many other smaller computers in an "Integrated Computing
Network" (ICN). This network also provides central and
distributed nodes for file management and storage,
graphical and printed output, network and production
control, and distributed computing.

The characteristics of some of the major applications in the ICN and the performance of the various Cray
computers on the Los Alamos workload are discussed, and
some experiences in creating the ICN and plans for the
future are presented.

INTRODUCTION

The Los Alamos National Laboratory is operated by
the University of California for the United States
Department of Energy (DOE). The Laboratory employs about
7000 people, and its mission is to perform research and
development activities related to national security and
energy programs.

° Talk presented by B.L. Buzbee.

To support its own and other national programs, the Laboratory has developed a state-of-the-art scientific computing network that supports over 6000 users (approximately 4000 are Laboratory employees and 2000 are users at other installations throughout the United States). The network, called the Integrated Computing Network (ICN), is shown conceptually in Figure 1. Approximately 150 computers are connected to the ICN, and it is separated physically or logically into three security partitions: Secure for classified computing, Open for unclassified computing, and Administrative for administrative computing.

At the center of the network are the large "worker" computers that today include two Cray Research, Inc. (CRI) Cray X-MPs, four Cray-1s, four Control Data Corporation (CDC) 7600s, one CDC CYBER 176 that is owned by the Defense Nuclear Agency (DNA), three CDC CYBER 825s,

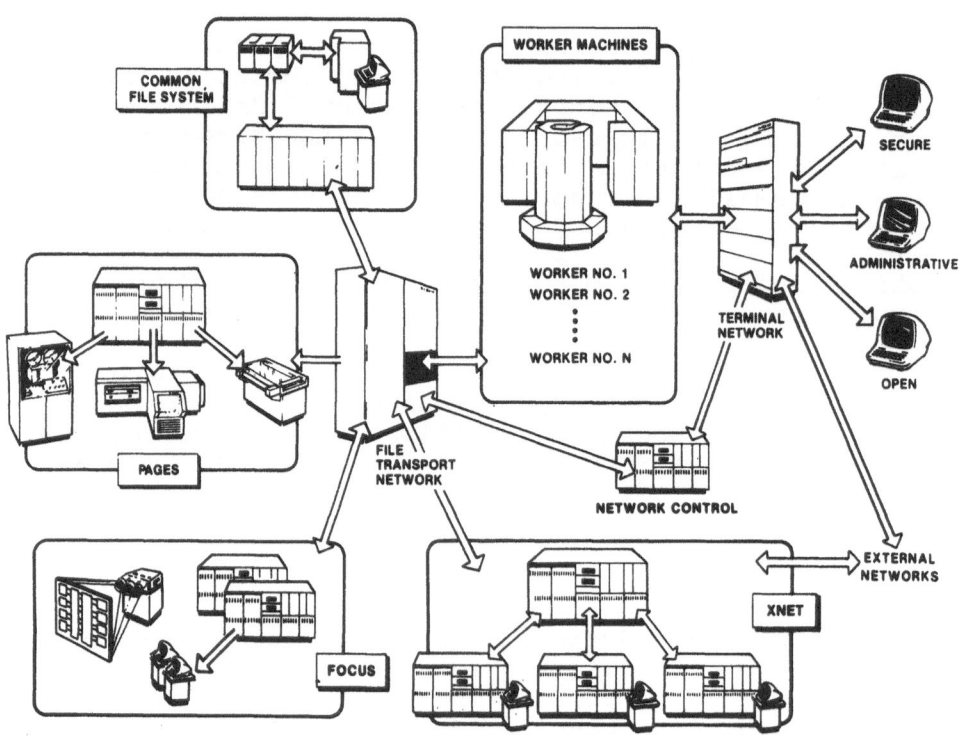

Fig. 1. ICN functional configuration.

and several Digital Equipment Corporation (DEC) VAX-11/780 computers. All of these computers run interactive time-sharing operating systems, and a common set of software is supported across all workers to allow users to move easily from one computer to another. The operating systems in use on the computers are:

CTSS (Cray-1s, Cray X-MPs). The Cray Timesharing System developed by the National Magnetic Fusion Energy Computing Center, Livermore, California, for the Cray-1. CTSS is the successor to LTSS and supports large-scale scientific and engineering computing. Los Alamos modified CTSS to make it operational on the dual processor X-MPs.

LTSS (CDC 7600s). The Livermore Time-Sharing System developed at Lawrence Livermore National Laboratory that supports scientific and engineering computing.

NOS (CDC CYBER 825s and 176). The Network Operating System developed by CDC that supports Computer-Aided Design/Computer-Aided Manufacturing (CAD/CAM) services, scientific, engineering,and administrative computing.

VMS (DEC VAX-11/780s). The Virtual Memory System developed by DEC for the VAX family of computers that supports scientific and engineering computing, database applications, office automation, and distributed computing functions.

UNIX (DEC VAX-11/780s). An operating system developed by Bell Laboratories that is used primarily for text processing, artificial intelligence research, and some scientific processing.

The common software available on all worker computers includes Fortran 77 compilers; three graphics libraries - our own Common Graphics System (CGS), the National Center for Atmospheric Research (NCAR) library, and DISSPLA; our Common Los Alamos Mathematical Software (CLAMS), a standard mathematics library (based on the SLATEC library, which was developed jointly with other DOE, Department of Defense laboratories and others); and common utilities to perform such functions as file shipping, graphics file processing, intermachine process-to-process communication, and production job submission.

To provide additional computing for users at remote locations or users who have a need for some local computing capacity, the Laboratory selected the DEC VAX as its standard distributed processor. There are currently about 35 VAX distributed processors in the Laboratory

and at remote locations connected to the ICN Extended
Network (XNET) through three other VAXes that act as
gateways into the network. All of the common software
available on the workers is also available on the distributed processors, and a "terminal passthrough" capability
has also been added to allow VAX local terminals to
access the worker computers.

Users typically access the network through terminals
in their offices. Table I gives the number and types of
terminals primarily used by the 4000 Laboratory computer
users.

Most of these terminals are connected to our
"terminal network",which is composed of about 25 DEC
PDP/11 minicomputers and 4 MICOM switches that act as
concentrators and switches to allow any user to access
any worker computer provided the user has authorization
to do so. Once a user selects a computer and has
provided the proper security and accounting information,
this part of the network becomes transparent and it
appears that the user is logged on directly to a worker
computer. Most terminals operate between 300 and 9600
bit/s, with about 100 Tektronix graphics terminals running at speeds up to 300 kbit/s. The IBM PCs and other
selected personal computers may perform terminal operations as well as file shipping within this part of the
network.

To enable large blocks of information to be trans-
mitted between worker computers and service nodes, the
file transport network was constructed with five System
Engineering Laboratory (SEL) 32/55 computers. Using
Los Alamos-designed interfaces and communication soft-

Table I. Quantity and Terminal Types

Approximate Quantity	Terminal Family
1200	TI 700s & 800s
1300	DEC VT100
500	Tektronix 4000
600	IBM PC

ware, this portion of the network allows files to be shipped at up to 50 Mbit/s.

A centralized node called the Common File System (CFS) is provided for long-term file storage. The system is accessible from any worker computer or distributed processor in the network and appears as an extension of a computer's local file storage system. It is based on IBM hardware with two 4341s and a 3083 as controllers, numerous 3350 and 3380 disks, and a 3850 mass storage system. Online capacity is about 3.7×10^{12} bits, although the users currently have stored about 1,400,000 active files with a total volume of about 18×10^{12} bits. Three storage hierarchies are supported: disk, online tape cartridge, and offline tape cartridge. The system runs Los Alamos-developed software and automatically migrates user files between the three tiers of storage. Files that are expected to be accessed frequently are stored on disk; infrequently accessed files are stored off line. Because of this balancing, response is excellent, as indicated in Table II.

Printed and graphical output is handled centrally by the Print and Graphics Express Station (PAGES). PAGES is a network node to which any worker or distributed processor may ship text or graphical files for processing. PAGES is controlled by two DEC VAX-11/780s that drive three Xerox 9700 laser printers (10,000 lines/min each), four FR-80 COM recorders (16-mm, 35-mm, and 105-mm film formats), a Dicomed D48 COM recorder (35-mm film), a CalComp plotter, and two Versatec hardcopy plotters. PAGES currently processes more than 2×10^9 bytes of information daily, which contributed to a total output of 3 million pages of print per month, 6 million

Table II. CFS Storage Hierarchy

Type of Storage	Total Information Stored (%)	Total Files Accessed (%)	Response Time
Disk (3350, 3380)	2	90	4 seconds
Mass Storage System (3850)	14	9	1.5 minutes
Offline	84	1	5 minutes

frames of microfiche per month, and about 250,000 frames of 16- and 35-mm film each month.

The newest network node, the Facility for Operations Control and Utilization Statistics (FOCUS), was implemented to control portions of the network, provide a unified operator interface, and gather utilization statistics. The FOCUS system consists of two VAX-11/780s, which operate the production computing environments on the Crays and the CDC 7600s. Production (long-running) jobs are submitted to FOCUS, which schedules and monitors the execution of the jobs and the worker machine status. FOCUS also provides a single operator station for all of the computers under its control, so that typically only one to two operators are required to operate the entire set of Crays and CDC 7600s.

To prepare for parallel processing, we have also installed a Denelcor HEP computer in the ICN. We are currently using this computer to conduct research into parallel processing algorithms and are making good progress. Based on early experiences, we expect to be able to parallel process our major application algorithms.

The major computing network resources available at Los Alamos are summarized in Table III and are described in detail in [1].

To support our computing hardware, networking, data

Table III. Major Computing Resources

Quantity	Description	Operating System	Total power (CDC 6600 = 1)
2	Cray X-MP	CTSS	80
4	Cray-1	CTSS	64
4	CDC-7600	LTSS	16
1	CDC CYBER 176 (DNA)	NOS	4
3	CDC CYBER 825	NOS	3
35	DEC VAX-11/780 (DPs)	VMS/UNIX	12
1	Common File System (CFS)	MVS	
1	Output Station (PAGES)	VMS/UNIX	
1	Production Controller (FOCUS)	VMS	
1	Denelcor Experimental Machine (HEP)	HEPOS	

Table IV. Operating Expenses

Category	$M	Percent
People-related expenses	19.9	44
Computer lease and maintenance	15.6	34
Material and supplies	10.1	22
	$45.6M	100%

communications, software, and general computing related services, we currently employ about 250 people in eight groups in the Los Alamos Computing and Communications Division. Our total 1984 fiscal year (FY84) operating cost for this part of the Division is expected to be about $45.6M. The major expenses incurred in FY84 can be categorized as shown in Table IV.

To recover these expenses, we charge all of our users for their use of the services. For FY84, our major computer usage rates are indicated in Table V.

Users also pay for their use of the network (monthly fee, connect time, traffic), storage on the CFS (storage and access), and output on the PAGES system (volume).

Table V. Major Computer Usage Rates

Machine	Cost ($/hour)		
	Day	Night	Weekend
Cray-1	594	297	148.50
CDC 7600	488	244	122
CDC CYBER 825	424	212	106
VAX 11/780*	30/mo	30/mo	30/mo

* This is a flat monthly fee per user. Hourly fees are not charged.

Due to the amount of automation and the efficiency of delivery of services at Los Alamos, the economies of scale are evident in our charging rates.

APPLICATION PROGRAM CHARACTERISTICS

Even with the powerful computing resources available at Los Alamos, the computing facility operates 24 hours/day, 7 days/week, at nearly saturated conditions.

Interactive timesharing systems are run on all computers, but there are two different types of loads on

Table VI. Representative Daytime CTSS Workload and Performance at Los Alamos

Type of Cray	1A	1S	1S	X-MP
Size of Main Memory (words)	1M	2M	4M	4M
Number of users logged on	24	18	27	37
Fraction of CPU to users	70%	83%	81%	87%
Fraction of CPU to system (about a third rechargeable)	18%	16%	18%	12%[*]
Response to small jobs (milliseconds)	480	290	330	240
Memory utilization	69%	78%	82%	79%
Number of user programs in core	4.5	6.8	11.7	9.4
Program loads/minute	155	125	160	230
Average size of program loaded (k words)	45	53	55	63
Disk controller utilization (per controller)	28%	30%	26%	NA
User disk I/O (Megawords/minute)	3	3	4	NA
Total disk I/O (Megawords/minute)	21	19	26	NA

[*] Per CPU
NA - Not available currently on the X-MP

the computers. During the night and on weekends, work-
loads on the supercomputers tend to be very large
scientific "production" programs that typically run for
several hours. During the weekday periods, the large
computers typically have about 30-40 users signed on
working on graphics and short debug or production runs
(up to about 10 minutes of CPU time). Production programs
are always waiting to be executed, even during the day,
so system performance tends to be very good.

Recent performance measurement work on the Cray-1 at
Los Alamos [2] is summarized in Tables VI and VII for
three memory configurations of 1, 2, and 4 million words of

Table VII. Representative Nighttime CTSS Workload and
Performance at Los Alamos

Type of Cray	1A	1S	1S	X-MP
Size of Main Memory (words)	1M	2M	4M	4M
Number of users logged on	2	2	4	6
Fraction of CPU to users	78%	87%	87%	92%
Fraction of CPU to system (about a third rechargeable)	10%	11%	11%	7%[*]
Response to small jobs (milliseconds)	150	140	120	150
Memory utilization	65%	75%	76%	80%
Number of user programs in core	3.3	5.4	7.3	7.7
Program loads/minute	55	75	70	95
Average size of program loaded (k words)	58	58	62	73
Disk controller utilization (per controller)	12%	15%	16%	NA
User disk I/O (Megawords/minute)	2	2	3	NA
Total disk I/O (Megawords/minute)	9	11	12	NA

[*] Per CPU
NA - Not available currently on the X-MP

central memory and for a Cray X-MP. Table VI reviews the typical daytime "timesharing" period, and Table VII reviews the typical nighttime "production" period.

As is seen in Tables VI and VII, CPU utilization of the Crays at Los Alamos is very high, with about 70-90% of the CPU delivered to the user programs. The programs that run on the Crays range from a few thousand words to over 3 million words of memory. Total traffic to the disks is about 10-30 Mwords/min, which is typically spread across four disk controllers onto 8 DD-29 or 16 DD-19 disk units.

To better understand the character of the existing programs, to suggest enhancements to programs and the software environment, and to provide information for our

Table VIII. Measured Characteristics of Codes 1-5 on a
Cray-1

Program	Millions of Floating-Point Operations	CPU Time (in seconds)	Millions of Floating-Point Operations /s (MFLOPS)	Average Vector Length	Percentage Vector
Program 1					
Problem 1	219.6	69.6	3.2	59	1.8
Problem 2	137.9	38.4	3.6	11	26.5
Problem 3a	5.5	1.9	2.9	8	29.1
Problem 3b	9.4	2.8	3.4	13	28.7
Problem 3c	16.9	4.5	3.8	21	28.4
Program 2					
Problem 1	4432.8	232.6	19.1	32	98.6
Problem 2	8595.6	4314.5	2.0	31	10.7
Program 3					
Problem 1	4215.2	902.2	4.7	26	36.3
Problem 2	1121.9	225.3	5.0	53	49.0
Program 4	136.3	69.8	2.0	20	77.7
Program 5	2427.1	85.5	28.4	63	69.5

high-performance computing research activities, we have also initiated a "Workload Characterization" project [3]. This activity has identified a representative set of five major production programs in use at the Laboratory, instrumented those codes, and measured their perform- ance, as indicated in Table VIII [4]. The first three programs are weapons design programs (hydrodynamics) and consume about 50% of the production computing resource. They typically have over 100,000 lines of Fortran and use dynamic memory allocation because of their large memory requirements. Programs 4 and 5 are particle-in- cell (PIC) programs of about 30,000 lines of Fortran that have been heavily vectorized.

As Table V shows, there is a wide range of perform- ance for these production codes and a large variation in the percentage of the program spent in vector operations. The Cray-1s can operate at over 100 MFLOPS, but the best sustained rate for a production code is about 30 MFLOPS. It should be noted that these programs run about 5-10 times as fast as they did on the previous supercomputer in use at the Laboratory, the CDC 7600.

FUTURE REQUIREMENTS

As we look to the future, our users are projecting requirements for capacity increases of about a factor of 6 by 1990. Translated into Cray-1 equivalents, this would mean that current Los Alamos users would consume about 65 of today's Cray-1s in 1990, if the projections hold. In addition to capacity increases, the users require capability increases as well, to be able to treat increasingly difficult problems. To continue to achieve overnight turnaround on these more difficult problems, users project that by the end of the decade they also will have requirements for computers that are 200 times faster (200x) and with much larger memories than those currently available. To realize 200x computers, advances in three major areas of computer architecture are re- quired:

1. Components - much faster components will be required to drive the cycle time lower (for example, the Cray-1 cycle time is 12.5 ns, and the X-MP cycle time is 9.5 ns).
2. Packaging - shorter "wires" will lead to more speed.
3. Parallel processing - the ability to apply more than one processor to the solution of a single problem.

209

By making various assumptions and by combining performance factors in manners that may (or may not) hold in
reality, it is possible to conceive of architectures that
would lead to 200x performance by about 1990. (For
example, reduce the cycle time to 1-2 ns, apply 16-32
processors, and if the performance factors are multipli-
cative, a peak performance improvement of approximately
100-200x might be seen.) It should also be noted that
more powerful machines will require much larger memories,
peripheral storage devices, and network Input/Output
(I/O) devices to remain balanced.

No matter how they are constructed, current and
future supercomputers require tremendous amounts of feeding, care, and cleaning-up after. This has serious
implications for both the users of supercomputer services
as well as for the providers of these services. From
the user's point of view, it would be preferable to not
have to do any program conversions (they would prefer to
do science or engineering rather than think about vectorization of parallelization). They would prefer to be
able to have new software tools that would provide more
insight into the real problems being solved (graphics,
higher level languages, expert system assists, etc.).
They will need very rapid interactive access to the
supercomputer for debugging, graphics, etc. Additionally,
tremendous amounts of short- and long-term storage will
be required as will rapid access to that information.
The usual conflicting user requirements (state-of-the-
art tools versus "everything is always changing," higher
level languages versus "but my program is 250,000 lines
of Fortran," high performance versus "but I don't want
to think in parallel") also will have to be considered.

AN INTEGRATED PLAN

To help satisfy requirements like those given above,
we believe that an integrated approach to providing
services works well for large or small installations.
As supercomputers are installed, the rest of the support
structure must remain balanced.

As a point of reference, the Los Alamos long-range
requirements and major hardware acquisition plan is given
in Table IX. Entries in the table indicate a planned
expansion of a particular part of the ICN. The equipment
base for early FY84 to which the entries in Table IX will
be <u>added</u> (generally) is given in Table III. Network up-
grades are continual.

Table IX. Long-Range Hardware Requirements/Acquisition
Plan

	Total User Requirements (Cray-1 equivalents)	Additional Worker Computers	Total Installed Capacity (Cray-1 equivalents)	CFS Upgrades	PAGES Upgrades
FY84	12	Cray X-MP	10	$2.4M	
FY85	15	Cray-1	13		$1.5M
FY86	19	Class VII*	17	$4.2M	
FY87	26	Class VII	26		
FY88	35	Class VII	33	$4.0M	$1.5M
FY89	48	Class VII	40		
FY90	65	Class VIII	47	$3.0M	

*Today's Cray-1, Cray X-MP, and CDC CYBER 205 are Class VI computers. Class VII computers are expected to be significantly more powerful (5-10x) than Class VI computers. The Class VIII is expected to be 100-200x more powerful than the Class VI.

Examples taken from Los Alamos' experience and future projections are given below. The examples are given in terms of feeding, caring for, and cleaning-up after supercomputers.

Feeding

To feed computers capable of processing several hundred million floating-point operations per second (MFLOPS) requires the ability to connect users to the computers at very high data rates and to be able to provide input files to the computers as required.

At Los Alamos, interactive timesharing operating systems are provided on all supercomputers. Because the machines are capable of processing such large quantities of information so rapidly, interactive communications requirements (particularly for graphics) are more severe than for other computers. The output side of these requirements is described in the "cleaning-up" section. Since humans operate at about the same rate on small and supercomputers, interactive input requirements are not substantially different for supercomputers than for other computers.

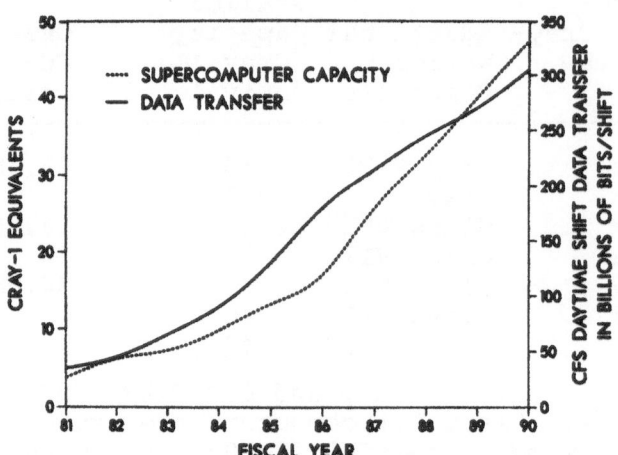

Fig. 2. CFS prime shift data transfer compared to supercomputer capacity.

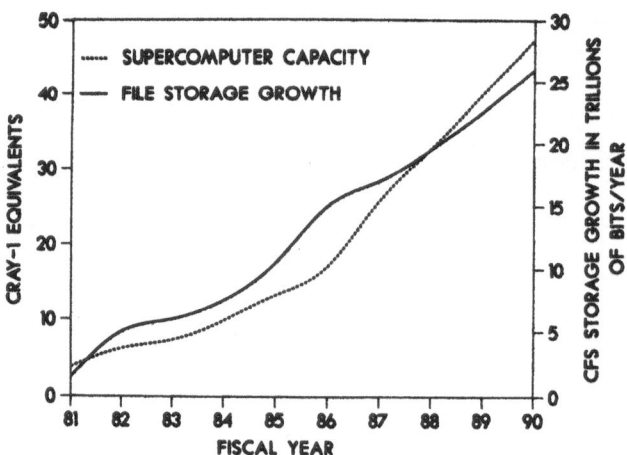

Fig. 3. CFS storage growth compared to supercomputer
capacity.

However, production runs on supercomputers also require (and produce) tremendous amounts of data, typically in the form of files. A typical production run on a Los Alamos Cray-1 uses several input files with a total file size of about 10 Mbytes. Three generic types of intermediate files are created with a total size of about 30 Mbytes, and three types of output files are typically saved for the long term. These files are typically sized between 200-400 Mbytes. Because this volume of information typically cannot reside permanently on a supercomputer, file storage systems such as the Los Alamos CFS are almost a necessity. The network must also be able to handle these very large blocks of information rapidly and accurately. With the Los Alamos CFS and network, users are satisfied if they can access 99% of their active files within one minute over the network; they are not satisfied with two-minute response or with bottlenecks at the start and end of the day. Figure 2 compares the data transferred from the CFS during day-time shifts with the installed computing capacity at Los Alamos. Note that the data transfer increases with the capacity of the computing facility.

The amount of information produced by supercomputers and stored for the long term also tracks very closely with the installed supercomputer capacity, in our experience. Figure 3 shows this trend for the past four years and our projections for future network storage requirements.

Caring For

To properly care for a supercomputer requires a long-term, active commitment. Users typically would like the complete set of services for a supercomputer that includes: (1) operation; (2) documentation - online, printed, and help packages; (3) education; (4) consulting - from program bugs to vectorization to numerical techniques; (5) programming support; (6) high-speed mathematics libraries; (7) high-quality graphics packages - 3D, color; and (8) a complete set of software tools including vectorizing compilers, interactive debuggers, screen editors, etc. The software investment on supercomputers typically far exceeds the hardware investment.

From an operational point of view, a centralized control center like FOCUS has tremendous payoff. It provides a consistent production control interface across several computers, unifies the operator interface, and does a better job of scheduling and load leveling than humans can do. With the FOCUS system and the other automated functions, it would be as easy to operate 20 Cray-1s and 16 CDC 7600s as it is to operate our current set. Feeding and clean-up, of course, are another matter.

Today's supercomputers are more reliable than yesterday's and hopefully less reliable than tomorrow's. We believe (and see) that machines such as the Cray-1 system should be capable of running at least 100 hours (four days) between <u>any</u> interruptions that are due to hardware problems that take the system down (these range from disk failures through logic problems on the computer). CDC 7600s historically run about 40 hours (two days) and VAXes fail about once a month.

Cleaning-up

As previously noted, today's supercomputers generate a tremendous amount of data, and the larger memory machines of the future will generate even more. Much of the output is in the form of interactive graphics, which poses very high bandwidth requirements on the network. Consider the graphical output problem for supercomputers. Users would like to be able to run "movies" of their numerical simulations on interactive graphics terminals. Each frame of a "movie" might require 2,000 by 2,000 bits for picture resolution, and each pixel might require 8-24 (say 12 is reasonable) bits of color information. Thus each frame might require 48×10^6 bits of information. To create a movie might require 8-30 (say 8) frames per second for a total throughput of 384 Mbit/s per user. A typical supercomputer might serve 30-40 simultaneous users, 5-10 of whom might be interested in high-performance graphics at any instant. Clearly, even with distributed applications, high bandwidth terminal connections will be required.

Supercomputer users also generate a tremendous amount of printed and graphical hardcopy output. Once again, our experience is that output generation correlates with the installed capacity, as shown in Figure 4, which compares the amount of output processed by our PAGES system to the installed capacity.

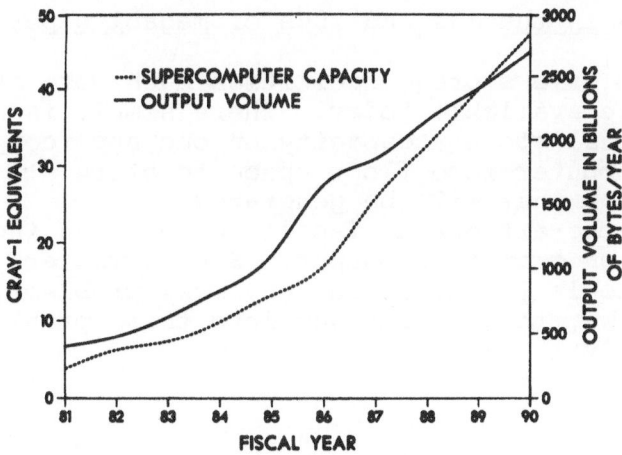

Fig. 4. PAGES output volume compared to supercomputer
capacity.

SUMMARY

Operating and providing supercomputer facilities
is a challenging job. Our experience has been that a
careful job of managing, planning, and executing plans
helps ease these tasks. We have had good success in
distributing the file storage, output, and operator
control functions in the network so that many worker
computers may be served by these functional modes. These
automation projects and provision of better user tools
have allowed us to reduce our operations staff from 110
people four years ago to 60 people today, while our
computing capacity increased by a factor of 4 during the
same period. These and other efficiencies are evident
in the charges for our services.

To provide adequate supercomputer services in the
future will require an integrated environment that should
include:

1. Very high-speed networks to deliver information to the
   users, to allow real distributed computing to take
   place, and to provide information that will be consumed
   by the supercomputers. It is reasonable to assume
   that individual users may require linkages to super-

computers in the 10s and 100s of megabits per second ranges.
2. Long-term file storage facilities much more capable than those available today. There simply is not enough local storage capacity on one supercomputer or enough computer-room floor space to store all of the information that will be generated.
3. State-of-the-art output facilities with an increasing emphasis on graphical output. Supercomputers are used as tools to solve real physical problems, and the best way to gain insight into these problems is graphically.

ACKNOWLEDGMENTS

The work described in this paper was performed under the auspices of the United States Department of Energy. The people of the Computing and Communications Division and numerous others are responsible for the achievements reported in this paper. Special thanks are due to Judith Valentine for her expert word processing skills and to Diana Tuggle and Shirley Cartwright for creating the figures.

REFERENCES

1. "Computing Division Two-Year Operational Plan, FY 1984-1985", Los Alamos National Laboratory report LA-9978-MS (January 1984).
2. W. Alexander, "1984 Workload and Performance Measurement on the Cray/CTSS Systems at Los Alamos", Los Alamos National Laboratory, unclassified report LA-UR-84-1610 (June 1984).
3. I.Y. Bucher and J.L. Martin, "Methodology for Characterizing a Scientific Workload", Los Alamos National Laboratory, unclassified report LA-UR-82-1771 (October 1982).
4. J.L. Martin, I.Y. Bucher,and T.T. Warnock, "Workload Characterization for Vector Computers: Tools and Techniques", Los Alamos National Laboratory, unclassified report LA-UR-83-305 (August 1983).

V. INDEXES

AUTHOR INDEX

Adelantado, M., 131, 141
Aho, V., 30, 32
Alder, B.J., 161, 174
Alexander, W., 207, 216
Arnold, C.P., 123, 141

Barkai, D., 145, 154,
    156, 157
Batson, P.E., 178, 197
Baudet, G.M., 123, 140
Baudoin, C., 73, 110
Berger, Ph., 115, 122,
    123, 127, 132,
    140, 141
Bernutat-Buchmann, U.,
    44, 67, 69, 70
Binder, K., 154, 157
Bohm, D., 176, 197
Boisseau, J.P., 134,
    135, 141
Bossavit, A., 73
Bourdel, F., 135, 141
Brent, R.P., 98, 110
Brittin, W.E., 180, 197
Brosens, F., 175, 180,
    182, 183, 196,
    197
Brouaye, P., 122, 140
Bucher, I.Y., 209, 216
Buzbee, B.L., 199

Calahan, D.A., 107, 110
Ceperley, D.M., 161, 174
Chappell, W.E., 180, 197
Chelikowsky, J.R., 163,
    174
Chen, C.H., 178, 197

Chen, S.S., 16, 31
Cohen, M.H., 159, 161, 174
Cohen, M.L., 163, 165, 174
Comte, D., 115, 120, 131,
    140, 141
Cosnuau, A., 135, 141

Davis, 63
Detert, U., 13, 25, 30,
    32
Devreese, J.T., 159-161,
    163, 170, 174,
    175, 180, 182,
    183, 196, 197
Dewe, M.B., 123, 141
Dickson, L.J., 122, 140
Dijkstra, E.W., 73, 110
Donahue, J., 165, 174
Donnet, S., 131, 141
Drouffe, J-M., 151, 157
Dubois, P.F., 107, 110

Ehlich, H., 45, 51, 70
Ehrenreich, H., 197
Eisenberger, P., 178, 197
Enselme, M., 134, 135,
    137, 141, 142
Ewald, R.H., 199

Feilmeier, M., 70
Feng, 39
Ferrell, R.A., 179, 197
Fields, J.R., 178, 197
Fitch, J., 151, 157
Flynn, 39
Fraboul, Ch., 115, 135,
    137, 139, 141, 142

SUBJECT INDEX